HYDRAULIC AND OPERATIONAL PERFORMANCE OF IRRIGATION SCHEMES IN VIEW OF WATER SAVING AND SUSTAINABILITY

SUGAR ESTATES AND COMMUNITY MANAGED SCHEMES IN ETHIOPIA

Zeleke Agide Dejen

Thesis committee

Promotor

Em. Prof. Dr E. Schultz
Emeritus Professor of Land and Water Development
UNESCO-IHE Institute for Water Education
Delft, the Netherlands

Co-promotors

Dr Seleshi Bekele Awulachew
Inter-Regional Advisor on National Sustainable Development Strategies
Water, Energy and Capacity Building Branch
Division for Sustainable Development, UN-DESA
United Nations
New York, USA

Dr Laszlo Hayde
Senior Lecturer in Irrigation Engineering
UNESCO-IHE Institute for Water Education
Delft, the Netherlands

Other members

Prof. Dr P.J.G.J. Hellegers, Wageningen University
Prof. Dr N.C. van de Giesen, Delft University of Technology
Dr M. Ayana, Arba Minch University, Ethiopia
Dr F.W.M. van Steenbergen, MetaMeta Research, 's Hertogenbosch

This research was conducted under the auspices of the SENSE Research School for Socio Economic and Natural Sciences of the Environment

HYDRAULIC AND OPERATIONAL PERFORMANCE OF IRRIGATION SCHEMES IN VIEW OF WATER SAVING AND SUSTAINABILITY

SUGAR ESTATES AND COMMUNITY MANAGED SCHEMES IN ETHIOPIA

Thesis

submitted in fulfilment of the requirements of
the Academic Board of Wageningen University and
the Academic Board of UNESCO-IHE Institute for Water Education
for the degree of doctor
to be defended in public
on Thursday 22 January 2015 at 2 p.m.
in Delft, the Netherlands

by

Zeleke Agide Dejen
Born in Harar, Ethiopia

CRC Press/Balkema is an imprint of the Taylor & Francis Group, an informa business

Published by
CRC Press/Balkema
PO Box 11320, 2301 EH Leiden, The Netherlands
e-mail: Pub.NL@taylorandfrancis.com
www.crcpress.com - www.taylorandfrancis.com

ISBN 978-1-138-02767-1 (Taylor & Francis Group)
ISBN 978-94-6257-169-3 (Wageningen University)

TABLE OF CONTENTS

DEDICATION

This thesis work is dedicated to my mother and father, who have worked hard throughout their life in a small village, raised me up, educated me, and hoped to see me as a man in a position to make a bit of a difference. I love you Mom! I love you Dad! I am now a man you wished me to be.

ACKNOWLEDGEMENTS

Clearly PhD research takes a significant portion of one's life. As such it obviously needs motivation, commitment and the courage to be able to tackle challenges in the due course. The journey is both difficult and exciting. It is difficult in that the time is long and requires patience and continuous hard work without being frustrated by a series of challenges. It is exciting because it paves the way and gives the opportunity to meet, communicate and work with highly inspiring and great people in the academia and beyond.

This PhD thesis is a result of continuous support and guidance of great people whose encouragements and advices have been speechless.

Primarily, I am extraordinarily grateful to my esteemed promoter Em. Prof. dr. ir. E. Schultz. I am really honoured being his PhD student. He has always been supportive and encouraging to me. He has been my strength and has always made me enjoy the challenges that I encountered on the way. He also visited my study areas in Ethiopia in January 2013, which was a great pleasure to me. What an honour to work with him! Without his keen support, intelligent guidance and valuable advices, this research could not have been realized.

I am very much thankful to my supervisor Laszlo Hayde, PhD, MSc, for his continuous support and supervision. He also visited my study areas in Ethiopia with Prof. Bart Schultz. Moreover, he made all the field equipments available to me whenever required; including current meters, EC-meters, divers, soil moisture sensors and GPS.

I am indebted to F.X. Suryadi PhD, MSc, for his valuable support in this research. He has been available to give me his expert advices and support when needed.

I am very much thankful to my local supervisor Seleshi Bekele, Dr. Ing, for his support and advice particularly while I was working in Ethiopia, which contributed to the success of this thesis.

My gratitude extends to Mr. Birhanu Mulu, BSc, Head Civil Engineering Department at Metahara Sugar Estate Scheme; Mr. Tibebu Mamo, BSc, Plantation Supervisor at Metahara Sugar Estate Scheme; and Mr. Abera Girma, MSc, BSc, Head Civil Engineering Department at Wonjj-Shoa Sugar Estate Scheme for their valuable assistance during my fieldwork.

I extend my wordless thanks to my mother and father for raising me up in a small village and educating me paying all the sacrifice where need be. You did all one can imagine for my success. I thank you and love you Mom and Dad!

My sincere gratitude also goes to my dear brothers Ketema, Yohannes, Awlachew and Kibebew and my dear sisters Addis, Genet, Hiwot and Mimi for their huge and endless love and support. I am proud to have you all beside me all along my life. I thank you!

Last but not least, I thank my wife Bezawit Azene for her thorough encouragements and continuous support. Thank you for all your patience and love!

SUMMARY

The two finite resources for irrigated agriculture, land and water, are globally shrinking and the pressure on these resources is increasing continuously. For the same reason, the rate of expansion in the global area of irrigated land has greatly decelerated during the last two decades. With limited freshwater and land resources, and ever increasing competition for these resources, irrigated agriculture (largest consumer of global freshwater resources) needs to improve its utilization of theme resources. There is a general consensus that the rate of increase in irrigation water withdrawal will go on slowing down in the next few decades. As such, the largest proportion of the required increase in agricultural production would have to be realized from already irrigated (cultivated) lands without major increase in the volume of water withdrawal.

Ethiopia is a least developed country in the Horn of Africa, with a total land area of 1.13 million km^2. The cultivable land of the country is estimated to be 72 million ha; while recent reports indicate that only about 25% (15 million ha) has been put under cultivation. Agriculture directly supports about 85% of Ethiopia's population and constitutes more than 80% of export values. However, the agriculture sector until recently remained under-developed and low productive. The sector is dominated by small-scale farming of for subsistence, facing challenges of poorly developed farming technologies, land degradation and large dependence on rainfall. The vast majority (95%) of the Ethiopian agriculture is rainfed, depending on rainfall of high temporal and spatial variability. Incidence of recurrent rainfall failures (droughts) aggravated by the effects of climate change had been occurring for several decades leaving a considerable number of people in the lowlands (low rainfall regions) at risk of food insecurity. The water resources of Ethiopia are enormous; amounting to about 125 BCM (billion cubic meters) of annual surface water potential and an estimated annual groundwater potential of 2.6 BCM. Total irrigable land potential in Ethiopia is estimated to be 5.3 million ha, including from surface water, groundwater and rain water harvesting. Equipped irrigated area to date covers only 700,000 ha, including schemes of all scales. The figure shows that the irrigated area covers only 12% of the potential and 5% of the cultivated land.

Ethiopia has currently embarked to an accelerated irrigation development plan, in which the irrigated land is planned to be increased to three folds in five years. Apparently, expansion of irrigated land through new irrigation developments is relevant in Ethiopia in view of its underutilized potentials of land and water. However, ensuring sustainability of the existing schemes is equally vital, which is clearly overlooked in Ethiopia. The majority of operational irrigation schemes in the country are characterized by a poor level of technical, hydraulic, operational and service delivery performance. Shortcomings include inadequate irrigation scheduling, inadequate operation plan, waterlogging and salinization, lack of adequate institutional setups for management, inadequate physical water control facilities, canal sedimentation and lack of adequate maintenance, lack of appropriate asset management, etc. Some of these challenges are critical to small-scale community managed schemes, while others are fundamental to large-scale schemes.

This PhD research concerns the performance of two large-scale and two community managed irrigation schemes in Ethiopia. The large-scale schemes are known as Wonji-Shoa and Metahara, while Golgota and Wedecha are the community managed schemes. Wonji-Shoa and Metahara were developed in the 1950's and 1960's, and irrigate areas of 6,000 and 11,500 ha respectively. They are located in the Awash River Valley, within the Central Rift Valley of Ethiopia, at about 100 km apart with Metahara

Scheme on the downstream. These are both public irrigation schemes growing exclusively sugarcane using Awash River as a source of irrigation water. Water is supplied to Winjo-Shoa Scheme from a pumping station on the bank of the river, and is distributed by gravity. Metahara Scheme is supplied with water by two gravity diversion structures on Awash River. Golgota is one of the community managed schemes considered in this study, and is located in between Wonji-Shoa and Metahara schemes, in the same river basin. A temporary gabion diversion structure supplies water to this scheme with a nominal command area of 600 ha. Wedecha community managed scheme is situated in the Central Highlands of Ethiopia, also in Awash River Basin, and has a nominal command area of 360 ha. Irrigation water is supplied from a reservoir created with a small embankment dam across Wedecha Stream, a small tributary of Awash River.

This PhD research aimed to evaluate the hydraulic and water delivery performance in the large-scale schemes with the objectives of evaluating the existing operation rules and proposing alternative options for more effective operation, sustainability and water saving. On the other hand, the research has carried out a comparative performance assessment and utility evaluation of internal irrigation service in the two community managed schemes.

Awash River Basin is the most utilized river basin in Ethiopia for irrigation. It is a basin of high socio-economic importance owing to its route of traverse in the driest north eastern Rift Valley Region. The river is the only source of water for over 5 million pastoralists and semi-pastoralists, and their cattle in the region. In addition, it is a source of municipal water supply for several towns along the river. Recently, there have been a number of large and medium-scale irrigation developments underway in the basin. Moreover, there are large numbers of community managed irrigation schemes under construction in an effort by the government to enhance food security by way of transforming the vast pastoral community to a semi-pastoral one. As such, competition for water in the basin has intensified in recent years, and there has been an increasing pressure on the existing schemes to improve water use efficiency. Obviously, increasing water demands in the basin would reduce the water share to the existing schemes. This in turn calls for a more effective irrigation water management that ensures better operational efficiency, adequacy and equity. Wonji-Shoa and Metahara schemes are among the major irrigation schemes in the basin, which need to address their water management. Community managed schemes, like the two schemes considered in this study, play a major role for food security and in alleviating rural poverty. In order to ensure the sustainability of these schemes, irrigation service, water productivity and institutional aspects for water management need to be addressed.

Performance assessment in irrigation and drainage is a systematic observation and interpretation of the management of irrigation and drainage schemes, with the objective of ensuring that the input of resources, operational schedules, intended outputs and required actions proceed as planned. The overall purpose of performance evaluation is to ensure improvement. Performance assessment has been a widely studied subject and concern during the last two decades within the context of diminishing land and water resources and the need to increase productivity of existing irrigation schemes. Accordingly, quite a large number of researchers have studied and addressed various aspects of performance of irrigation schemes around the world. However, there were almost no irrigation performance evaluation initiatives in Ethiopia in the past, particularly for the large and medium-scale schemes. In this research, performance issues identified in the large-scale schemes are hydraulic (water delivery), water saving and related environmental issues of waterlogging and salinization. On the other hand, for community managed schemes the performance issues identified are related to irrigation service provision, water and land productivity, and institutional aspects for water management.

Manual operation of flow control structures at Wonji-Shoa and Metahara schemes is not only labour intensive and laborious, but also ineffective. The complex hydrodynamic behaviour of the systems is not well understood, and the existing operation takes little account of these effects. The hydraulic performance was first evaluated at each scheme from routinely measured flows at offtakes classified as head, middle and tail. Canals of 9 and 11 km length, with 16 and 15 offtakes respectively were considered in Wonji-Shoa and Metahara schemes. Flows at the offtakes were measured two times a day over three months (January, February and March) for the years 2012 and 2013. These are the months of low flow in Awash River, during which water availability is minimum. Adequacy (relative delivery), offtake delivery efficiency, equity and reliability were used as indicators of the water delivery performance. Moreover, a comparison of annual irrigation water diversion versus demand was made based on measured diverted discharges using stage-discharge relationships established in the head of the main canals of the schemes.

Water delivery performance from routinely collected offtake flow data at the two large-scale irrigation schemes has clearly shown the status of the water delivery performance. Routine flow data measurement is time consuming and cumbersome. However, one can monitor the actual conditions and the results are more reliable. As a first step water diversions measured at the canal heads and calculated demands were compared, and significant over supply was found for each. Then, for selected (main and secondary) canals, offtake discharges were monitored for two years over the three dry months to use for determination of the water delivery performance indicators.

The annual measured water diversions exceed annual demand by 51 and 24% at Wonji-Shoa and Metahara schemes respectively. Results of water delivery performance evaluation indicated that unlike the classical assumption that head offtakes deliver larger supplies, at both of these schemes, tail reach offtakes were supplied with excess and more water under the existing operation. It was shown that water delivery to head offtakes is acceptably right at both schemes due to reasonably small fluctuation in water levels in the parent canal in the head reach. Canal emptying and filling had the worst effects on the hydraulic performance (particularly adequacy) of middle reach offtakes at Metahara Scheme. However, offtake relative delivery (adequacy) decreased from head to tail for Wonji-Shoa Scheme. Tail excess supplies were due to inadequate operation and hyper-proportional nature of the combination of offtake and water level regulating structures. Overall, offtake delivery equity levels under the existing operation at both the schemes were fairly acceptable, although there were significant fluctuations from one year to another. This was because the portion of overall diverted excess water that was lost within a tertiary unit at both the schemes was relatively small compared to losses off-farm (distribution), operational losses at main system level and tail runoff. Offtake flow measurements and efficiency indicators have indicated that field percolation losses account for only about 20 and 10% of the total annual excess diversion at Wonji-Shoa and Metahara schemes respectively. The remaining amounts percolate from the distribution systems and are drained to drains within the schemes and at the saline tail ends.

It was observed that the operational losses resulting from continuous fluctuation of the flow, nature of the flow control structures and sudden closure of tail offtakes let huge amounts of excess water to the tail ends. As such, flow monitoring has revealed that the irrigation supply to an irrigation block called 'North block' (900 ha) in Metahara Scheme, located at the tail end, was more than twice of the demand. The average groundwater level in this irrigation block is about 1 m below the ground surface, while sugarcane requires a groundwater table of 1.5 to 2.0 m below the soil surface for optimal growth. This has been a threat to the sustainability of the scheme. Soil salinity (ECe) in the irrigation block ranged between 1.5 and 3.5 dS/m, which is moderately high to adversely affect cane growth. At this salinity level, the average reduction in crop yield for sugarcane is about 10%.

Irrigation supply to the lower (tail) quarter portion of Wonji-Shoa Scheme was nearly as high as 200% of the demand in some fields. Shallow groundwater tables resulting from ill-water management has been an increasing challenge at this scheme as well. The offtake relative delivery is nearly 1.0 at the head reaches and 1.3 at the tail reaches with an average excess delivery of 30% at the tail, while the total annual excess is 51%. Although the proportion of excess diversion, actually delivered to the offtakes is small on average (about 20%), water distribution had the worst effects at the tail ends. This is so because of inundation and poor drainage in the downstream ends. Over an area of about 1,000 ha, groundwater levels has risen to less than 1 m below the soil surface, which will worsen rapidly without major changes in water management. Modification of operational measures for water management can significantly improve the water delivery performance and adverse effects. Some of these measures based on field observation, measurement and evaluation have been recommended in this thesis.

Hydrodynamic simulation models are useful tools to understand the complex hydrodynamics of canal irrigation systems and to evaluate their hydraulic performance. The effects of different operational interventions on the hydrodynamics and the resulting performance can be evaluated. These models have been used by several researchers for evaluation of irrigation performance or to aid improvement in operation. However, the application of these models so far have in large focussed in irrigation schemes shared by individual or groups of water users. In this case, a hydraulic model has been used in the public sugar estate irrigation scheme of Metahara with no individual water users. DUFLOW, a one dimensional hydrodynamic model, was calibrated and used to assess the existing operation in terms of water delivery (hydraulic) performance of the scheme. Measured discharges at 16 offtakes along the canal and flow depths measured at two locations (1+300 and 7+100) in the canal system considered were used for calibration. The model was also validated using measured offtake discharges under a different hydrodynamic condition than for calibration. Discharges for calibration were measured with current meters and flow depths with divers (pressure sensors) installed at the two locations. Chezy roughness coefficient (C) and discharge coefficients of structures (C_d) were used as calibration parameters. For setting up the model, canal bed profile and cross sections were surveyed with Total Station surveying equipment. Detailed data on the location and features of structures were also surveyed by a walk through survey. The hydraulic performance under the current operation as well as operation scenarios that would enhance operational efficiency, equity and save irrigation water were simulated by the model.

Further on to the evaluation from routinely monitored flow data, hydraulic simulation at Metahara Scheme enabled a better understanding of its hydrodynamics and the water delivery performance under the existing operation. Simulation has resulted in an annual excess water diversion of 41 Mm^3 (million cubic meters), which is 27% of the annual demand. The simulated excess closely matches with the excess diversion, which was determined from routine flow measurement (37 Mm^3). Simulations have also shown that daily canal filling and emptying causes more fluctuation in water levels in the middle reach than in the head and tail reaches for the existing operation. Hence, the maximum fluctuations in offtake water delivery were observed in the middle reaches. There was a rapid depletion in water levels in two to three hours after offtake opening. The hydraulic sensitivity of the structures in the middle reach and inadequate operation of the reservoir outlet and water level regulators were the main causes. On the other hand, offtake discharges in the head reach generally went on gradually increasing due to an increase in water levels in the parent canal during irrigation hours in this reach. Flow rates to tail offtakes, however, remained more or less the same during irrigation hours.

Overall, it was found that the amount of water lost at tertiary and field levels was only about 7% of the surplus water for Metahara Scheme. Seepage in the main and secondary canals was relatively small due to reduced infiltration by clogging by fine river sediments. The simulation showed that more than 50% of the excess diversion was discharged at the downstream ends of the system, where severe waterlogging and salinization were evident. Regarding the offtake adequacy of the supply, the tail reach offtakes were supplied with a relative delivery (Allen *et al.*) amount of 1.17 on average (17% excess). Once these offtakes have been closed water runs to the tail swamps downstream of the offtakes. The head and middle offtakes had relative deliveries of 1.05 and 0.84 respectively. The average operational efficiency for the head, middle and tail offtakes determined from the simulated offtake flows for the existing operation were 0.93, 0.94 and 0.85 respectively, which all perform good. Overall equity of delivery to offtakes along the canal system for the existing operation was determined to have a spatial coefficient of variation (CV) of 0.15, which can be regarded as 'fair'. As such, the hydraulic performance shortcomings revealed from the hydraulic simulation of the existing operation of Metahara Scheme were: 1. excess water diversion; 2. tail runoff that resulted in waterlogging threatening sustainability; 3. under supply of middle reach offtakes and over supply of tail offtakes.

Three different operational scenarios that would enhance equity, adequacy and also save water were simulated, and the effect of each scenario on the hydraulic performance was evaluated. The scenarios were: 1. adopting 24 hours irrigation with modified settings of structures and steady flow in the system; 2. adopting 12 hours irrigation with modified settings of offtakes; 3. adopting 9 hours irrigation with modified settings for operation of the main intake, reservoir and offtakes. These operation scenarios would have annual water savings of 15, 11 and 14% of the demand respectively, which are significant savings for a gravity surface irrigation scheme. The operational efficiency was determined to be higher than 0.9 in each scenario for the simulated offtake flows. Similarly overall offtake delivery equity determined based on the effective and supplied (simulated) offtake flows was fairly adequate (CV between 0.06 and 0.12). The performance ratio (relative delivery) for scenarios 1 and 2 in each reach indicated a performance of 'good' as per the standard. However, for Scenario 3, its performance was 'fair' for the head and middle reaches, while it was 'good' for the tail reach.

Comparative irrigation performance assessment enables comparison between schemes and within the same scheme over time as a means of tracking changes. Cross comparison in irrigation schemes helps to compare outputs from irrigation and bulk impacts of agricultural systems. External indicators basically provide limited information on the internal processes of the irrigation system. In comparative performance evaluation it is not the actual numerical value of the indicator which is important, but the relative performance of the agricultural system in relation to other schemes. While internal (process) performance assessment is mainly concerned with the achievement of internal management targets such as flow rate and timing of water delivery, comparative evaluation gives insight on how productive and efficient land and water resources are used for agriculture. Golgota and Wedecha community managed schemes in this study were evaluated with three groups of comparative indicators; namely, water supply, agricultural production and physical sustainability.

These two schemes vary in several aspects including source of water, method of water acquisition, water management, size of landholding, etc. At Golgota Scheme, water is relatively not a scarce resource, and farmers are responsible for all aspects of water management without involvement of a government agency at all. Moreover, the water users use water for free except for their own routine maintenance. However, at Wedecha Scheme decision on water diversion from the source is being made by an external government agency, while farmers are responsible for their water sharing and on-farm water management. Farmers of Wedecha Scheme pay an irrigation water fee of 48 US$/ha per year to the agency. With these differences, comparative evaluation was made to examine the utilization of land and water resources and irrigation sustainability. The two groups of indicators for comparison (water supply and agricultural production) proposed by the International Water Management Institute (IWMI), to which a third group called physical sustainability indicators was added were used.

The comparative evaluation of performance showed that there is a significant difference in resource utilization of the schemes. At Golgota Scheme, where all aspects of water management are the responsibility of water users, the annual relative water supply was more than twice of Wedecha Scheme. Institutional aspects for water acquisition and irrigation water fee were identified as key factors for an efficient use of water in these schemes. Although participatory irrigation management at Wedecha Scheme has resulted in a reduced water diversion, its supply schedule and degree of reliability had their own impact on productivity. Water productivity at Golgota Scheme was relatively inferior compared to Wedecha Scheme. Owing to the present more scarce water at Wedecha Scheme, this seems right. However, on the other hand, extremely low water productivity of Golgota Scheme is a concern even under generous water supply. Annual land productivity at Golgota Scheme was found to be nearly twice of that of Wedecha Scheme. However, land productivity is not a function of the water availability alone, but also of other factors such as soil type, use of fertilizers, crop varieties, etc. It was also found that water availability can affect land productivity indirectly. It was determined that readily water availability at Golgota Scheme increased willingness of farmers to invest more on their piece of land and it also enhanced increased irrigation intensity, which all increased output per unit of irrigated land. Annual irrigated land productivity at Golgota Scheme (as high as 6,400 US$/ha) is significantly high compared to similar schemes in Ethiopia and the average in Sub-Saharan Africa. High irrigation intensity (about 250%) contributes the largest share for high land productivity. The annual output per unit harvested area was 2,600 US$/ha at Golgoata Scheme, while the average at Wedecha Scheme was 1,970 US$/ha.

Physical sustainability as an indicator was meant for sustainability of the irrigated areas and for utilization of the design irrigable lands at the two community managed schemes. Both irrigation ratio and sustainability of irrigated land were higher for Golgota Scheme. Water management was the main factor. Self water management by the water users and absence of irrigation water fee were the main reasons identified for expansion of irrigated land at Golgota Scheme.

The comparative assessment of performance in the two schemes identified the following key issues: 1. farmers are willing to pay for a minimum routine maintenance of their own, but not to an external agency for water management; 2. willingness of farmers to invest on their piece of land and hence land productivity depend on arrangements for irrigation water management; 3. for such smallholder farmers, the larger the land holding size, the higher is the land productivity due to the willingness of the farmers to use inputs and to spend full time working on their piece of land; 4. The suitability of irrigation water management arrangements depends on the type and condition of the water source.

A reasonable irrigation water fee at an appropriate rate was concluded to be a useful water saving incentive for enhancing water productivity. While all aspects of water management (operation, scheduling, water sharing, conflict resolution, routine maintenance, etc.) are best done by the water users association, interventions by an external agency particularly in flow measurement and monitoring is recommended under the existing conditions. Moreover, it was identified that collective volumetric assessment (water fee policy) works much better for effective water use than the current area-based fee as being practised at Wedecha Scheme.

Internal (process) irrigation performance evaluation is meant to assess the internal processes like the flow rate of water delivery, its timing, its duration, dependability of the supply, etc. The rationale for internal irrigation performance evaluation is to improve the irrigation services to water users. Evaluation of internal indicators generally requires measured quantitative data on water deliveries. However, service-oriented irrigation water measurement is generally given little or no priority in small-scale schemes, particularly in least developed countries, and such data are therefore generally not available. Thus, flow related data would need to be collected from the schemes whenever need arises. However, internal performance evaluation in these schemes from measured flow data in the field would not well address the needs of smallholder farmers. This is because in community managed schemes with poor off-farm and on-farm irrigation infrastructure, water users generally have different and several criteria for evaluation of the irrigation services, which the conventional methods do not address. Hence, a different approach of evaluation based on farmers' perceptions is an alternative.

Irrigation service level (utility) can be evaluated from the perspective of the water users (main stakeholders in the business) based on their qualitative responses on water deliveries. In Golgota and Wedecha schemes, there were no data on irrigation deliveries and their timing to evaluate service delivery performance to each group of water users. The utility of irrigation service was hence evaluated from qualitative data collected from sampled water users at different locations within the schemes. Three utility factors; namely tractability, timing and dependability were used and each factor was decomposed into two utility sub-factors. The sub-factors considered were stream size and point of water delivery for tractability; time of water arrival and duration of delivery for timing; and knowledge of future delivery and certainty of availability for dependability. Perceptions of sample farmers on the importance and suitability of each utility sub-factor were collected using a questionnaire survey at each scheme. Fuzzy set theory was used to represent and aggregate the attitudes of the water users at the head, middle and tail reaches. The aggregated qualitative expressions of farmers were then converted to a numerical indicator of service levels (utility) ranging from zero to one.

Results of the utility analysis were determined both for the importance of the factors and the suitability of the service relative to the utility factors. At Golgota Scheme, tractability was the most important factor, while dependability was the least important. Farmers were more concerned about the flow rate and point of water delivery than its dependability. On the other hand, at both sub-systems of Wedecha Scheme dependability was the most important factor, while timing was the least important, which indicates that they were more concerned about the certainty and reliability of water availability. At Golgota Scheme, the overall aggregated utility was higher for the middle reach, while it was the same and lower in the head and tail reaches. On the other hand, at Wedecha Scheme, overall utility decreased from the head to the tail reaches.

The purpose for improved utility is to improve irrigation service and hence increase productivity. Thus, average agricultural output in each reach was determined at each scheme in order to see any relation with utility. Results indicated that on average there appeared no relationship between utility and output at Golgota Scheme due to more uniform and better utility values across the reaches. Moreover, agricultural output is also a function of several other elements of the agricultural system which were not taken into consideration. However, at Wedecha Scheme, average agricultural output steadily decreased from the head to tail reach the same way as the utility did. Issues identified to be fundamental for irrigation service quality (utility) are: institutional arrangement for water management; condition of water division and farm structures and their proper operation; and implementation capability of the water users associations.

The performance concerns in the large-scale and small-scale community managed irrigation schemes were identified and evaluated. For the large-scale schemes, excess water diversion (need for saving water), threats of rising groundwater levels (waterlogging) and salinization, and ineffectiveness of manual operation resulting in ill-hydraulic performance were the main concerns. The major threats to the sustainability of these schemes are salinization and shallow groundwater levels that resulted from excess water supply and low hydraulic performance. Nearly 1,000 ha of land at each scheme, mainly in the tail reaches, were under a threat of shallow saline groundwater tables. These schemes do not have sub-surface drainage system for controlling groundwater levels. Controlled water diversion and improving the hydraulic performance through adequate operation not only saves substantial fresh water, but also reduces the risk of further waterlogging and ensures sustainability. For the small-scale schemes major performance concerns were related to the institutional arrangements for water management, irrigation service delivery, land and water productivity, and off-farm and on-farm water management. As these are schemes for smallholder farmers, productivity (both land and water) are crucial to them. The sustainability of these schemes is dependent on sustainability of institutional setups for adequate water management, operation and maintenance, and reliability of irrigation service.

While it is possible to make a comparison between large-scale irrigation schemes and small-scale community managed schemes, it is also important to note the basic differences. In fact, there is a basic difference in typology between these schemes. The large-scale schemes are public schemes with mono cropping (sugarcane) and there are no individual water users. Thus, water management issues related to water sharing among farmers and internal irrigation service do not apply. Sugarcane is an annual crop with more or less uniform water requirement throughout. In the community managed schemes, farmers practice several kinds of cropping patterns for intensification. Due to more or less similar water management and agricultural practices at the two large-scale schemes, water and land productivity remained almost similar. Land productivity (output per ha based on net revenue) from Sugarcane without processing is about 425 US$/ha/year, which is only 1/8 of the land productivity of Golgota Scheme and a quarter of that of Wedecha Scheme. However, processing Sugarcane to sugar increased the annual net land productivity by 550%. Net water productivity of the large-scale schemes from Sugarcane is about 0.018 US$/m^3, while that of the community-managed schemes ranges between 0.1 and 0.3 US$/m^3 for supplied irrigation water. Processing Sugarcane to sugar, in fact, significantly increased water productivity as in the case of land productivity.

Adequate water management for irrigated agriculture in general holds a considerable significance for the future of the Ethiopian agriculture. The short-term irrigation development plans of the country show that small-scale irrigation schemes are considered the major suppliers of food, while large-scale irrigation developments are largely planned for government owned large-scale agro processing industries, mainly sugar. In fact, the importance of small-scale irrigation schemes for food security in Ethiopia can be well recognized due to the demographic and land ownership situation.

Small-scale farming (irrigated and rainfed) currently provide more than 95% of the food production. However, there seems a need for transforming the agriculture system in Ethiopia by moving away from subsistence to medium and large-scale irrigated agriculture for sustainable food security. This is also important for the sector to play its share in the development plans of the country. As such development of medium and large-scale irrigation potentials in the lowlands would need to be accelerated for production of food crops. Meanwhile, the overall performance and sustainable management of the developed irrigation schemes deserve an equal consideration. In this regard, capability of institutional setups, adequacy of physical asset management, adequacy of maintenance, sound irrigation scheduling, service oriented management and reliability would be major issues to be well integrated into the water resources management policy of the country and implementing stakeholders.

1. INTRODUCTION

1.1   General

With steady increase of the global population, the contribution of irrigation towards boosting agricultural production is enormous. Particularly, in some emerging and least developed countries irrigation development and use is a backbone to the extent that it is responsible for the nations' welfare and feeding the vast majority of their population. In these countries, an increase in production of 100-400% is being attained by irrigation, which depicts the importance of irrigation to agricultural production (Food and Agriculture Organization of the United Nations (FAO), 2005). According to International Fund for Agricultural Development (IFAD) (2014) and Hess (2010) only 20% of the world's total croplands are irrigated. However, these lands contribute to some 40% of the global agricultural harvest. The figure indicates that irrigated agriculture on average is roughly more than two and half times as productive as rainfed agriculture. Agriculture depending on rainfall has failed to produce enough food, and with increasing rainfall variability, productivity of rainfed agriculture is expected to diminish. To meet increasing demand for food by 2050 the global agricultural production would need to increase by 60% of the production in 2005 (FAO, 2012). As such without significant investments in irrigation, agricultural production is unlikely to cope with ever increasing demand for food.

It was identified that globally 60% of the diverted fresh water for agriculture does not contribute directly to food production. This amount of water is discharged because of poor water control, inefficient irrigation systems with leaky conveyance and distribution, poor on-farm water management practices, etc. (World Agriculture Forum (2009). It depicts that only about 40% global fresh water abstracted for irrigation is being effectively used for consumptive use in agriculture. Part of the amount of the discharged water of these systems is lost to saline groundwater or to poor quality drainage water. However, in some cases, discharged irrigation water can be recovered in the downstream reaches. Agriculture consuming about 80% of fresh water abstraction in several least developed courtiers is considered the most inefficient water user sector. With increasing number of countries facing water shortages, agriculture is expected to face a serious water stress in several regions. Thus, water scarcity remains to be a major challenge to feeding the global population. According to FAO (2013) by 2025, 1,800 million people are expected to be living in countries or regions with 'absolute' water scarcity, and two-thirds of the global population could be under 'stress' conditions. FAO (2005) also forecasts that without changes in efficiency of water use, by 2050 the world will need as much as 60% more water of the abstraction in 2005 for agriculture, which remains a challenge to the sector.

The rate of increase in water withdrawal for agriculture in the next few decades will not continue as was in the last three or four decades. FAO (2006) forecasts that the expected global agricultural water abstraction by 2030 would be about 14% higher than the abstraction in 2000. However, this figure is relatively low compared to the projected increase in irrigated area during the same period. This depicts that agriculture faces the challenge of producing more food with less water. It is expected that the discrepancy in water withdrawal and irrigated land expansion can be bridged by improvement in irrigation efficiency, thereby reduction in the withdrawals needed for irrigation water per irrigated land. Declining fresh water resources and increasing of competing water demands are among the main causes for slowing the rate of agricultural water withdrawal. This may also be aggravated by the impacts of climate change. On the other hand, it is expected that agricultural production from irrigated lands would need to increase by about 13% per decade during the next few decades to feed the increasing world population (FAO, 2003). Several studies show that the contribution of expanding agricultural land for the required increase in production is relatively low; and the contribution of cultivated lands, particularly that of irrigated lands will be much higher. Although there are various views on the speed of the required increase in agricultural production, the major part of the increase (80-90%) would by and large have to come from already cultivated lands; among other means, by improved irrigation practices, increased intensity, improved drainage practices, and increase in storage (Schultz et al., 2005).

Improving the performance of irrigation schemes through various interventions is considered a key issue for addressing the need for increased productivity of irrigated lands under pressure on water resources. Many irrigation schemes, particularly in least developed and emerging countries, are characterized by a low level of overall performance. The technical and economic performance of pubic irrigation schemes in these countries has generally been far below potential, and that of large-scale irrigation schemes in some cases is particularly very low (Darghouth, 2005). These schemes have been characterized by high unreliability of water supplies. However, large-scale irrigation schemes are generally shared by groups of water users and are often complex; hence require appropriate institutional setups and technical and operational plans for adequate performance. Areas of poor irrigation performance include mismatch of supplies and demands, insufficient maintenance, inadequate manual operation of structures, operational leakages and field losses, poor irrigation service, waterlogging and salinization. A large part of the low irrigation performance is, however, attributed to inadequate water management at scheme, system and field levels (Cakmak et al., 2004). As a result, in several irrigation schemes, irrigation water has been used at a very low efficiency, hydraulic performance has been low and irrigation service to farmers has been stumpy. The main causes for ill-performance were related to inadequate institutional setups and non-flexibility of the hardware of the schemes.

Many large-scale irrigation schemes in least developed countries are out of the reach of smallholder farmers. However, in these countries, small-scale irrigation is the major contributor to food security and improvements in rural livelihoods. Particularly in Sub-Saharan Africa, South and South Eastern Asia, small-scale irrigation schemes play a vital role for food security. There are several reasons for promoting small-scale irrigation schemes (Tafesse, 2003) including: lower investment costs, ease of maintenance, more flexible control of water by the users, possibility of reaching remote and poor farmers, possibility of water management by the water users, less negative environmental impacts. Nevertheless, many small-scale irrigation schemes in least developed countries are often characterized by low land and water productivity, poor physical infrastructure, less reliable water supply and unsustainable asset management. If these factors undermining the benefit of these schemes are addressed, small-scale irrigation schemes can play a big role for ensuring food security for the vast smallholder communities in these countries.

In view of the fact that water shortage will be a major constraint to agricultural production and that there is a need for increase in the productivity of irrigation schemes, the overall performance of schemes would have to improve. Water needs to be used more efficiently and water diversions per unit of irrigated land need to be reduced. With expected slowdown in expansion of irrigated land, greater focus seems to be put on improvement of existing irrigation schemes and their effective long-term operation and maintenance. Plusquellec (2009) stresses that, given diminishing fresh water resources and declining irrigation expansion, improving the productivity of existing irrigation schemes by addressing their deficiencies in management and poor performance in a holistic manner can no longer be ignored. Moreover, there is a need for intuitional transformation as has already been implemented in many irrigation schemes around the world. Appropriate mechanisms for saving irrigation water need to be implemented in schemes based on convenience.

This research concerns two large-scale sugar estate irrigation schemes and two community managed irrigation schemes in Ethiopia. The large-scale irrigation schemes are called Wonji-Shoa and Metahara, and are public irrigation schemes. The schemes grow excessively sugarcane and are operated by a public sugar enterprise. These schemes are manually operated gravity systems with extensive networks of open canals for water conveyance and distribution. The community managed irrigation schemes are named Golgota and Wedecha and are small-scale schemes. The key stakeholders of these schemes are smallholder farmers. The schemes are community managed, because farmers are responsible for management of irrigation water and maintenance of their infrastructure through their water users association (WUA).

The research aimed to evaluate the hydraulic and water delivery performance in the large-scale schemes with the objectives of evaluating the existing operation rules and proposing alternative options for more effective operation and water saving. In these schemes, the research made special emphasis on the hydraulic (water distribution) aspects of the performance. Moreover, the research also carried out a comparative irrigation performance assessment and irrigation service (utility) evaluation in the two community managed schemes. External (comparative) performance and irrigation service delivery to farmers were the major concerns of this research on these schemes.

1.2   Structure of the thesis

This thesis comprises of ten chapters, whose brief description is as follows. Chapter 1 highlights a general overview of water and irrigated agriculture in a global perspective. Chapter 2 gives a brief introduction of the schemes considered in this study, rationale of this study, objectives, and general research methodology. Chapter 3 highlights water and land resources, demography, economic condition, and levels of irrigation development with respect to potential in Ethiopia and gives a detailed description of the irrigation schemes considered along with their salient features on existing water management. Chapter 4 presents general concepts and guiding principles on comparative, irrigation service and hydraulic (water delivery) performance evaluation. Chapters 5 and 6 present the water delivery (hydraulic) performance assessment in temporal and spatial scales from routinely monitored flows at offtakes, for Wonj-Shoa and Metahara Sugar estates respectively.

Chapter 7 is about the hydraulic modelling (DUFLOW), used to evaluate the hydrodynamic behaviour of a canal system of Metahara Scheme under the existing operation rules. Moreover, an alternative operation rule for enhancing water delivery performance and water saving are presented. Chapter 8 presents the results of a comparative performance assessment in the two community managed schemes, a discussion of cross differences, and issues related to these differences. Chapter 9 presents an assessment of existing irrigation service levels (utility) in community managed schemes using qualitative data, with the fuzzy set approach. Chapter 10 shows an evaluation and the main performance challenges in large-scale and community managed schemes, relevance of each in the Ethiopian conditions and the lessons learnt in each case. It also presents the conclusions and the way forward.

2. BACKGROUND AND OBJECTIVES

Modern irrigation development started in Ethiopia in the 1950s in the Awash River Valley with the objective of producing industrial crops such as cotton and sugar cane and horticultural crops. In the 1960s, private irrigated agriculture (most of it owned by firms from overseas) expanded in all parts of the Awash River Valley and in the Lower Rift Valley of Ethiopia (Awulachew et al., 2007). Located in the upper Awash River Valley, Wonji-Shoa Sugar Estate irrigation scheme was constructed by the Dutch holding company, HVA, and the first phase of development for an area of 5,000 ha was completed in 1954. A few years later, the scheme was expanded with an intention of increasing sugar production. The total original irrigated area of the scheme for sugarcane was 6,000 ha. However, currently additional expansion pressurized irrigation development is underway at this scheme. Wonji-Shoa is the first modern irrigation scheme in Ethiopia.

Metahara Sugar Estate irrigation scheme is also located in the upper Awash River Valley about 100 km downstream of Wonjo-Shoa Scheme. Its establishment commenced in 1965 and was commissioned in 1968 by the same Dutch agricultural development company, after an agreement was signed between the then Ethiopian government and the company. Its irrigated area exclusively for sugarcane currently covers 11,500 ha of land. Water is distributed by gravity with an extensive network of open canal systems. The source of irrigation water for both schemes is Awash River. Following the 1974 Ethiopian revolution and conquer of the country by the military "Derg" regime, all irrigation schemes in the country belonging to private firms were nationalized. Wonji-Shoa and Metahara were among the irrigation schemes nationalized after the 1974 revolution.

The community managed irrigation schemes: Golgota and Wedecha are also located in Awash River Basin. Golgota is found in the upper Awash River Valley between Wonji-Shoa and Metahara Schemes, and draws its irrigation water from Awash River. A temporary still structure made of gabions is used to raise the level of Awash River and divert irrigation water to Golgota Scheme by gravity (Girma and Awulachew, 2007). The nominal command area of Golgota Scheme is 600 ha. Wedecha is situated in the Central Highlands of Ethiopia near Debre Zeit town at a distance of about 50 km to the East of Addis Ababa. It has two sub-systems called Godino and Gohaworki, each supplied with its own diversion from the main canal. The source of irrigation water is an embankment dam reservoir on Wedecha Stream, a small tributary of Awash River. The nominal command area of Wedecha Scheme (two sub-systems combined) is 360 ha.

Owing to its agro-climatic and demographic conditions, Awash River Basin, where the schemes of this study are located is a basin of huge socio-economic significance in Ethiopia. The river along with its tributaries directly supports more than 5 million people in the region for all kinds of water needs including agriculture, domestic, and livestock. In this regard, the existing large-scale irrigation schemes would have to implement a water saving strategy in their operation in order to ensure sustainable water utilization. The river flows in the semi-arid and arid region of Ethiopia characterized by high year round temperatures as high as 40 °C. Rainfall distribution in the basin is variable ranging between 1,500 to less than 200 mm per annual. Over 70% of the basin is semi-arid or arid with annual rainfall between 150 mm and 600 mm. In the basin, on average 70% of the annual rainfall is lost by evaporation (Taddese et al., 2007). In addition to the natural causes of water stress in the basin, the upper Awash Basin and its major tributaries have been subjected to major human-induced environmental stress. Demographic pressure on natural resources including deforestation in the basin to transform land uses to agricultural lands and for firewood has accelerated the situation.

2.1   Rationale of the study

Many large-scale irrigation schemes in least developed countries are typically manually-operated gravity systems, and their operation is often complex and monotonous. This requires continuous monitoring of structures and flows to make sure that the water delivery is demand responsive. Hydrodynamics of water distribution in such systems is also very complex and often not adequately understood by canal managers and operators. In irrigation systems where there is intermittent flow in the conveyance and distribution systems, the complexity of the hydrodynamics increases. This is because canal emptying, for example due to night time closure requires refilling, which could take significant time to reassume design water levels and discharges.

Water saving in the semi-arid Awash River Basin has recently become crucial due to expansion of large and small-scale irrigation developments in the basin. Excess irrigation water use in the large-scale schemes has already challenged their sustainability due to waterlogging and salinization at the tail end of the systems. Water conveyance and distribution in Wonji-Shoa and Metahara Schemes is based on non-continuous (intermittent) flow in the main systems. This means that field irrigation takes place only during day hours, while flow in the systems during night hours continues at a reduced rate to be stored in night storage reservoirs. Obviously, the flow in the main systems is either completely unsteady or quasy-steady as a result of alternate increase and decrease of the flows. Apparently, this requires effective operation rules that would minimize the impacts of the complex hydrodynamic behaviour on the water distribution, thereby enhancing equity and efficiency.

The two large-scale irrigation schemes are blamed for excessive irrigation water diversion, while the schemes still claim water stress during dry months of the year. In fact, it is not the amount of water diverted at the head that ultimately matters, but the way it is distributed and used within. To address these issues, it is imperative to well understand the hydrodynamic characteristics of the systems and to carry out detailed hydraulic analyses of the main water conveyance and distribution system. To the best of my knowledge, previous studies focusing on operational and water delivery/hydraulic performance evaluation in these schemes are absent. Based on the water delivery/hydraulic performance evaluation, alternative operation rules that would enhance water delivery equity, adequacy and saving need to be developed.

Small-scale subsistent irrigation is by far dominant in Ethiopia. These schemes play a vital role in improving the livelihoods of the smallholder farmers. However, existing small-scale community managed irrigation schemes face various problems related to operation and maintenance, water management and sustainability. These problems have greatly reduced their benefits and challenged their overall sustainability. The institutional arrangements for operation and maintenance and the day-to-day decision making process for water management in these schemes are different from one to another. As such, there is no single best model of water management that would ensure best results in terms of sustainability and enhancing water productivity.

So, a need arises to identify which arrangement for water management in community managed irrigation schemes functions better. For the Golgota and Wedecha schemes, the water diversion responsibilities and irrigation water fee situations are different. These in turn have clear implications on water productivity and long-term sustainability of the schemes. Comparative performance evaluation and internal irrigation service utility assessment would assist to identify the best operational water management arrangements in these irrigation schemes.

As a nation with least developed water resources, Ethiopia needs to speed up development of water resources projects for irrigation for the benefit of its people. However, sustainability and productivity of already developed schemes is equally important. The two key performance factors to consider in existing irrigation schemes are in fact the issue of sustainability, irrigation service and water productivity. Although these issues are vital, they received little attention both from the government and researchers. Particularly in Awash River Basin, water is becoming an increasingly scarce resource due to expansion of numerous water resources development projects for irrigation and domestic supplies. Moreover, the basin is located in the semi-arid Central and Eastern part of the country. Since the past few years, competition for water in the basin has specifically been rising as a result of large expansion of community irrigation schemes as means of enabling sedentary farming for the local pastoralist population and implementation of large-scale irrigation schemes for agro-processing. As such, the traditional water abstraction levels in the existing schemes will most probably be challenged within a few years from now. So, enhancing overall long-term sustainability and saving water by making appropriate management of it are concerns to be addressed in existing irrigation schemes.

2.2   Scope of the study

In the sugar estate large-scale irrigation schemes, this study concerns with hydraulic and water delivery performance, with special attention to evaluation of the current operation rules in terms of matching supply with demand, adequacy, equity, dependability and efficiency of water distribution and delivery to various parts of the systems. Moreover, the impacts of the complex hydrodynamic behaviour of the systems on water distribution and delivery were evaluated and understood. Based on the results of the evaluation, alternative operation rules that would enhance water delivery performance and water saving will be proposed. DUFLOW, a one dimensional hydrodynamic model, was calibrated and validated, and used as a tool to assist this evaluation and operation rule development for Metahara Scheme.

In the community managed schemes, this study made a comparative performance evaluation and irrigation service utility assessment. Selected relevant comparative (external) performance indicators were applied for comparison in terms of various criteria such as water productivity, land productivity, physical sustainability and, water/irrigation supply. Moreover, for each scheme the internal irrigation service delivery was assessed using the methodology of fuzzy set theory that enables evaluation of the level of service from qualitative data on irrigation water delivery. The study will propose institutional and operational water management arrangements that would ensure better irrigation service and sustainability in these schemes. These water management interventions could also be extended to other similar community managed schemes in Ethiopia.

The research will finally make a critical analysis of the issues related to irrigation water management and performance with respect to the large-scale and community managed schemes. It will make a distinction in the performance and sustainability challenges between the two categories of irrigation schemes.

2.3   Research questions

The Awash River Basin is the most utilized basin for irrigation in Ethiopia. The basin has an irrigation potential of 205,000 ha of land and extends through the Central to

North Eastern semi-arid region of the country. Six large and medium-scale irrigation schemes are located in this basin including Wonji-Shoa and Metahara. The basin is predominantly inhabited by pastoralist communities without permanent settlement. Recently, the government has been implementing several irrigation development schemes in the basin, with an attempt to transform these communities to semi-pastoralist. This in turn has increased the competition for water in the basin, and it is most probable that the water scarcity in the basin, particularly during the dry season, will be intensified in the years to come. With water becoming an increasingly scarce resource in the basin, Wonji-Shoa and Metahara schemes are blamed to have been diverting excess water supplies and inefficient use of water. Water management within the schemes is problematic and is characterized by high inequity and inefficiency levels. There have also been critical waterlogging and salinization problems, which might have been the result of over-irrigation and inefficient irrigation water management. There is a general consensus that the trends of water use in these schemes would not continue in the same way like what it has been in the past. Effective and sound system operation rules are believed to be potential water management interventions, which would ensure equity and efficiency of water distribution and water saving. Accordingly, the research questions which arose for these schemes are:

- how do the total water diversions into these schemes relate to demands and basin-wide water management?
- how does the existing operation and hydraulic features of structures affect the spatial and temporal water delivery performance (equity, adequacy, efficiency, reliability) in these schemes?
- will an alternative operation rule, based on a thorough hydrodynamic simulation, enhance effective operation, hydraulic performance and efforts in water saving at Metahara Scheme?

Community managed small-scale irrigation schemes, while playing significant roles in the efforts of the government to achieve food security, they are less sustainable. In many cases, they often fail to achieve the objectives they were designed and implemented for. Golgota and Wedecha community managed schemes are among such schemes. The institutional setups for irrigation water management and water distribution regulation in these schemes are different. The problems in these schemes include weak institutions, poor operation and maintenance guidelines, low water productivity and lack of long-term sustainability. The question arises as to which water management model and setup is best suited to enhance irrigation service delivery and sustainability of these schemes. Specifically:

- how does the external performance in these schemes under the existing water management setups compare with each other, and which key interventions are important for the difference?
- how does the internal irrigation service level in these schemes relate to institutional setups, water diversion entitlements and other water management viewpoints?
- which institutional setups, irrigation water fee policies and water management models would enhance water productivity and ensure overall sustainability in these schemes?

2.4   Research hypothesis

The hydrodynamic behaviour of traditional manually operated large-scale gravity irrigation systems is often less understood by operators and managers, while it plays a major role in water distribution and delivery. The situation worsens in unsteady flow systems with alternate canal filling and emptying, resulting in ineffective operation that causes significant water losses and inadequate water delivery performances. This study's hypothesis is that thorough investigation of the hydrodynamic behaviour of large-scale irrigation systems and consequent modification in operation enhances the water delivery performance and can save significant amount of water in these schemes. Small-scale community managed irrigation schemes, particularly in Sub-Saharan Africa, lack sound institutional and operational setups that could achieve increased productivity, reliable irrigation service, and long-term sustainability. This study puts forward that appropriate institutional, water acquisition and irrigation water fee setups greatly enhance irrigation service, productivity and sustainability in these schemes.

2.5   Research objectives

Enhancing hydraulic (water delivery) performance and saving irrigation water through effective operation are the key objectives to be achieved in this study for the large-scale irrigation schemes. In order to address the above, specific objectives are formulated as follows:

- to carry out thorough assessment and comparison of irrigation demand and water diversion trends in these schemes, and address issues related to future basin-wide water management;
- to evaluate water delivery performance indicators in spatial and temporal scales using routinely monitored offtake discharges under the existing operation rules, and identify the underlying major operational and hydraulic factors;
- to simulate the flow in the main water conveyance system of Metahara Scheme using a hydraulic model DUFLOW, and develop an alternative operational option (rule) that would enhance the hydraulic performance and water saving;

Improving irrigation service, land and water productivity and long-term sustainability through appropriate institutional and water management approaches are the main objectives of this study for the community managed schemes. Specific objectives are:

- to evaluate and rate the external performance of the two schemes with relevant indicators for comparison, and identify possible factors for difference under existing water management practices;
- to determine the internal service delivery level (also head-middle-tail) of performance within each scheme and its relationship with institutional, operational and other water management setups;
- to come up with and propose sound institutional and operational water management setups, which would better enhance irrigation service, water productivity and sustainability in the community managed schemes.

Finally, the research aims to compare and contrast the large-scale and the community managed irrigation schemes performances, and put forward the major differences and key issues in each case.

2.6   General research methodology

This study is largely based on extensive primary field data collection in the irrigation schemes. Primary data central to this study for the sugar estate schemes are canal profiles and cross sections, details of structures, data on existing operation rules, meteorological data, discharges, water levels, etc. On the other hand, key data to this study of the community managed schemes are existing institutional setups, irrigation service from farmers' perspective, discharges, agricultural produce, land holding size, etc. The research methodology employed in this research for the sugar estate schemes is as follows:

- flow monitoring at offtakes categorized as head, middle and tail reaches, with current-meters and v-notches;
- evaluation of water delivery adequacy (relative delivery), efficiency, equity and dependability from field measured flows;
- canal profile surveying (total station equipment), cross section survey (measuring tape and levelling) and a walk-trough detailed survey of structures for setting up the hydraulic model (DUFLOW);
- measurement of water levels at salient locations in the main canal (with divers) and discharges at several offtakes (current-meters) for model calibration and validation;
- setting up the DUFLOW model, calibration, validation and simulation of existing water management, and for development of adequate operation rules.

The research methodology followed at the community managed schemes is as follows:

- flow measurement with current-metering and stage-discharge relations in the main supply canals to assist in evaluation of water supply indicators;
- questionnaire survey of sampled water users (from head, middle and tail reaches) on the levels of irrigation service, water use, land holding size and agricultural produce;
- evaluation of external performance with selected performance criteria, and internal service delivery level with the fuzzy set approach from qualitative data;
- Development of an adequate water management setup for community managed schemes that would enhance irrigation service, water productivity and sustainability.
  Figure 2.1 shows a simplified conceptual framework of this research.

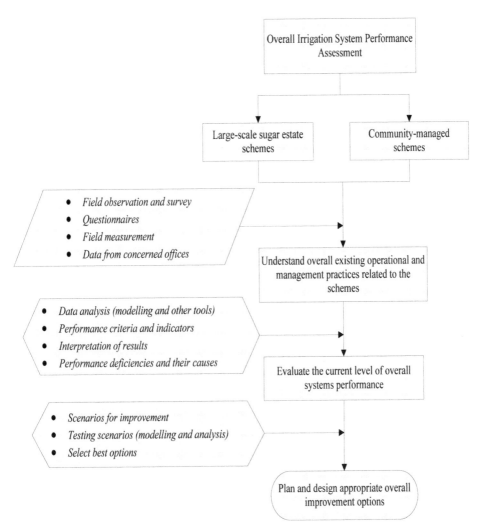

Figure 2.1. Simplified conceptual framework of the research

3. DESCRIPTION OF ETHIOPIA AND IRRIGATION SCHEMES OF THIS STUDY

3.1 Water and land resources

Ethiopia is a least developed and land locked country located in the horn of Africa. It has a total area of 1.1 million km^2, which fall under three major and distinct agro-climatic zones: cool humid (Dega), cool sub-humid (Woina dega) and Semi-arid (Kola). The country has 12 river basins, of which 3 are dry (ephemeral), basins characterized by storm flash floods. In terms of potential, the country has about 125 BCM (billion cubic meters) of annual renewable surface water and 'uncertain' 2.6 BCM of annual renewable groundwater resources. Of the 9 river basins which offer the surface water resources of the country, only two; namely, Awash River and rivers of the Rift Valley Lakes Basin flow within the country, while the other 7 rivers are transboundary. These 7 transboundary rivers including Blue Nile account for 95% of annual surface runoff. The annual per capita water availability including all the water resources draining out of the country is 1,560 m^3, which falls a little lower than a threshold value for water scarcity of 1,700 m^3, according to Falkenmark et al. (1989). With this substantial amount of surface runoff draining out of the country, the per capita water availability in Ethiopia is definitely among the countries with serious economic water scarcity. However, understandably the water scarcity in Ethiopia is economic, rather than physical.

Nearly 66% of the total land area is potentially suitable for agriculture, which is equivalent to 72 million ha (FAO, 2010). Nevertheless, due to various factors including climatic, demographic, socio-economic, etc., only about 25% of the total cultivable land was being put under cultivation by the same year. The Ethiopian highlands, constituting about 45% of the total land area, are regions facing high demographic pressure on land and water resources. On the other hand, the lowlands in the southern, south-eastern and south -western parts of the country, with sparse settlements, offer huge and unutilized land resources potentially suitable for agriculture. However, rainfed agriculture is least productive in these regions owing to little and the erratic nature of rainfall both in amount and distribution. Actually, very little irrigation infrastructure has been so far developed in these areas to bring these vast areas under irrigation.

3.2 Economic and demographic conditions

Agriculture is by far the backbone of Ethiopia's economy; like for many least developed countries. Agriculture in Ethiopia is largely characterized by a smallholder subsistence nature. It provides over 85% of the total employment, 43% of foreign exchange earnings, and approximately 50% of the GDP (Gross Domestic Product) (FAO, 2010). The vast majority (95%) of the share of agriculture for the GDP is produced by smallholder farmers cultivating less than 1 ha of land. Various efforts by the government are underway since 2007 to diversify the economy; e.g. to increase significantly the share of manufacturing sector in the economy. Among the efforts is also to develop the agricultural sector to move away from its subsistence nature through implementation of well designed agricultural policy and strategies.

The government of Ethiopia well understood that agriculture should play a primary role in the economy with a gradual shift and increase of the share of the manufacturing sector in the economy. For a country with such a vast population depending on agriculture, the industrial sector is highly dependent on the agricultural sector. National development policies cantered on these facts have enabled Ethiopia to have records of one of the fastest growing economies in Africa since 2004. Ethiopia is the second most populous country in Sub-Saharan Africa with a population of about 92 million. The economy has achieved a successive economic growth at an average annual growth rate of 10.6% between 2004 and 2013. The Gross Domestic Product (GDP) of the country at an official exchange rate was US$ 41.6 billion in 2012 and the population in the same year was 91.7 million (WorldBank, 2012). With a per capita GDP of US$ 453, Ethiopia is still among the least income countries in the world.

The Ethiopian highlands and Central Plateau (altitude higher than 1,400 m+MSL (mean sea level)) are the centre of the economic activity of the country, and carry about 90% of the country's population and 75% of livestock (FAO, 2010). Owing to the unsuitable semi-arid climatic conditions, the lowlands in the north-eastern, south-eastern and south-western parts of the country (altitude less than 1,400 m+MSL) are sparsely populated. These areas are inhabited by only about 10% of the population. Overall, about 80% of the total population lives in only 37% of the total area of the country, concentrated in the Central Plateau and northern highlands.

3.3   Water and land resources utilization for irrigation

Water and land resources development for beneficial uses in Ethiopia remained insignificant for centuries. The country lacked suitable institutional, political, technical and technological capability that would be able to utilize these resources for irrigated agriculture in a sustainable way. In terms of land and water resources, there is an estimated 5.3 million ha of potentially irrigable land. Of the total potential, 3.7 million ha is from surface water (small, medium and large scale), while the remaining 1.6 million ha is from rain water harvesting technologies and groundwater (Awulachew, 2010). In terms of utilization, only about 12% (about 640,000 ha) has been irrigated by 2010 (Ministry of Water Resources (MoWR), 2010). Regarding the overall consumptive use water development for irrigation, municipal and industrial uses, only 5% has been developed so far (World Water Assessment Programme, 2006). The total irrigated land accounts for only 5% of the total cultivated land of about 15.8 million ha. The discrepancy in irrigation potential and development clearly shows huge under development of the water resources. The remaining vast agricultural activity (over 95%) in Ethiopia is exclusively rainfed.

Typology of irrigation development in Ethiopia broadly falls under four different categories. These categories include traditional irrigation schemes, modern small-scale irrigation schemes, modern private irrigation schemes, and public irrigation schemes. Small-scale irrigation (traditional and modern) account for more than 70% of the total irrigated land in Ethiopia. These schemes belong to smallholder farmers with average landholding sizes of 0.25 to 0.5 ha (Awulachew et al., 2005). Smallholder farming (irrigated and rainfed) indeed dominates the agriculture in Ethiopia and are the major sources of food supply in the country. Though there has been large acceleration in irrigation expansion during the last decade, the country still lacks sound irrigation infrastructure that could well reach smallholder farmers. As such about 90% of the smallholder farmers do not have access to irrigation infrastructure. The irrigation potential of the country for large and medium-scale irrigation developments is little tapped. However, with large number of these schemes currently underway (pubic development projects and communal schemes), irrigated area under these schemes will escalate in the next few years. Table 3.1 highlights irrigation schemes typology.

Table 3.1. Irrigation schemes by typology in Ethiopia

S. No.	Typology	Area, %	Area, ha
1	Traditional small-scale irrigation schemes	39	250,000
2	Modern small-scale irrigation schemes	30	192,000
3	Modern private commercial irrigation	5	32,000
4	Medium and large-scale irrigation schemes	26	166,000
	Total	100	640,000

Recurrent droughts and dry spells resulting from large spatial and temporal variations of rainfall have been a major challenge in the Ethiopian agriculture, which left the country food self-insufficient for several decades. Irrigation development in Ethiopia is by far a key to address problems of food insecurity and ensure social welfare. Apparently, moving away the Ethiopian agriculture from its nature of subsistence and rainfall dependence to a resilient and sustainable one is inevitable to enhance national food security and alleviate the large rural poverty.

To this end, the Ethiopian government has been engaged in enormous accelerated water resources development for almost a decade now, through construction of large-scale water storage and irrigation facilities, and community-based irrigation developments. It was planned to boost the total irrigated land from 640 million ha in 2010 to 1.8 million ha at the end of 2014. Although there have been some unpredictable delays in some of the projects, with expected completion of several of the large irrigation projects combined with small-scale community schemes, the plan will by large be met.

3.4   Description of sugar estate irrigation schemes

3.4.1   History in brief

Wonji Shoa Large-scale Irrigation Scheme is located in the Awash River Basin at 8°21' to 8°29' N and 39°12' to 39°18' E. It is situated at about 100 km on the upstream of Metahara Scheme in the Central Rift Valley of Ethiopia. The average altitude of the area is 1,550 m+MSL and the mean annual rainfall is 831 mm. The scheme was developed by the Dutch agricultural development company, HVA, when first 5,000 ha of land was completed in 1954. Later the irrigated area was expanded to 7,000 ha in the early 1960's. Wonji-Shoa Scheme was the first commercial large-scale irrigation scheme in Ethiopia. Currently, the scheme has an irrigated area of 7,000, exclusively for sugarcane excluding recent pressurized irrigation expansion.

Metahara Large-scale Irrigation Scheme is also located in the Ethiopian Rift Valley between 8°21' to 8°29' N and 39°12' to 39°18' E. Situated at an altitude of 950 m+MSL, the area is semi-arid, with a mean annual rainfall of 543 mm. An agreement was made between the then Ethiopian government and the company for development of 10,000 ha of irrigated land for sugarcane and a sugar factory. Its development started in 1965 and was completed in 1968 by the same Dutch company. The whole development took 4 years, which included two diversion intakes, main water conveyance and distribution systems, reservoirs and all flow control structures. Currently, Metahara scheme has an irrigated area of 11,500 ha for sugarcane, and is one of the major large-scale irrigation schemes in the country. Both Wonji-Shoa and Metahara schemes were nationalized in 1974, and currently the schemes are public. Location of the two schemes is shown in Figure 3.1.

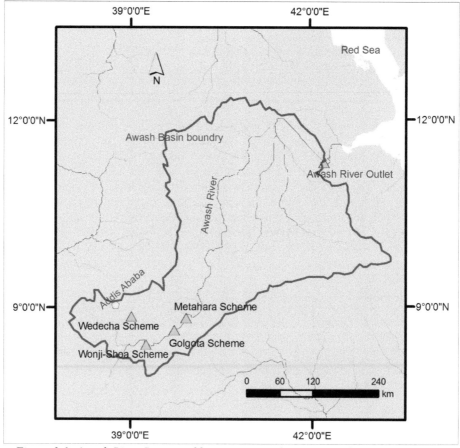

Figure 3.1. Awash River Basin and location map of irrigation schemes of this study

3.4.2 *Source and availability of irrigation water*

The source of water for both Wonji-Shoa and Metahara Large-scale irrigation schemes is the Awash River. The river originates from the highlands in the west of Addis Ababa and drains into interconnected lakes in the north-eastern border part of Ethiopia near Djibouti. It is one of the 12 major river basins of Ethiopia and traverses through a length of 1,200 km. The basin has a total drainage area of 112,700 km^2 and annual runoff of 4.6 BCM (Taddese et al., 2007). It is a perennial river and is the most utilized river basin in Ethiopia for irrigation. The river basin is sub-divided into 3 reaches; namely Upper, Middle and Lower Awash. The two sugar estates are located in the Upper Awash reach. There are large numbers of medium and small-scale irrigation schemes all along the river in addition to the two large-scale sugar estate schemes. Currently, over a 60,000 ha large-scale irrigation development in the downstream reach (Lower Awash), called 'Tendaho Scheme' is underway and is partly completed. Recently, there is also a large expansion of small-scale irrigation schemes underway by the government's efforts to reduce the vulnerability of the pastoralist community to recurrent rainfall failures (droughts). The river flows from the Central Plateaus through the driest North-Eastern semi-arid Rift Valley of Ethiopia. The agro-climatic condition of the region makes it a river of high socio-economic significance.

Awash River is a direct source of water for more than five million pastoralists and semi-pastoralists population in 'Oromia' and 'Afar' regions of Ethiopia. Moreover, the river also produces hydropower in its Upper reach (Koka Dam) and is a source of municipal water supply for several towns along the river. The release of water from the Koka reservoir is based on power generation and not on the demands on the downstream (Berhe et al., 2013). With ever increasing water demands of various sectors in the basin, Awash River is the most stressed river basins of Ethiopia.

For Wonji-Shoa Scheme, water is diverted with 8 parallel pumps by direct pumping from the river. As this scheme is located in the Upper reach of the river, the river stage fluctuation is within acceptable limits for direct pumping. Of the 8 pumps, 6 or 7 are operated at the same time based on field irrigation demand. Water is pumped into a settling basin from where it flows to the main canal and is distributed by gravity (Figure 3.2). There are 5 major night storage reservoirs within the system to store water during off-irrigation (night hours). There is little irrigation development on the upstream of this scheme, and there is relatively abundant water available for irrigation for the scheme. However, with ever increasing demands for water in the river basin, including industrial, municipal and irrigation, there is currently an increasing pressure on the scheme to improve its efficiency of irrigation water use.

a. Flow over weir at d/s end of selling basin   b. Main canal after the selling basin
Figure 3.2. Wonji-Shoa scheme main canal head reach

Water is diverted to Metahara Scheme by two diversion headworks with gated regulators, called 'Main intake and 'Abadir'. The main intake located about 3 km on the downstream of Abadir is a concrete diversion weir with sluice gates having 3 compartments that regulate the flow of water into a concrete lined canal of 10 m^3/s design capacity (Figure 3.3). This intake diverts water for 8,000 ha of land located on the right bank of the river. The current research focused only on the water management aspects of the main diversion (on the right bank). Abadir intake is a gabion (made of stone and wire mesh) weir on the upstream of the main intake, and it supplies water to about 3,500 ha of land located on the left bank. At Metahara Scheme, there happens a big fluctuation in the stage of the Awash River, particularly during dry seasons. There are a series of small and medium-scale irrigation schemes on the upstream of Metahara Scheme, which significantly reduce downstream flows. Moreover, there is a minimum flow of about 10 m^3/s to be let downstream of Metahara Scheme during the dry season. As such, there is generally a significant seasonal fluctuation in the river stage and hence irrigation water availability at Metahara Scheme.

a. Intake at Metahara Main diversion   *b. Main canal downstream of the intake*
Figure 3.3. Metahara Scheme Main intake and main canal head

3.4.3 Water management, flow control and field irrigation practices

Irrigation water is fully managed by the estates themselves. Water is being diverted, conveyed, distributed and applied to the irrigated fields by the employees of the estates. Both estates pay irrigation water fee to Awash Basin Authority (ABA) based on measured irrigation water flows at the heads of the main canals. Diverted volume of water at each estate is being measured using a stage-discharge relation established at the head reach of their main canals. Records of stages are kept on daily basis from the staff gauges, from which monthly diverted volumes are determined. Water is pumped more or less at a constant rate at Wonji-Shoa Scheme over 24 hours; while off-irrigation (night) flows at Metahara Scheme is being reduced to about 6 to 7 m^3/s. Night storage reservoirs distributed within the schemes balance the day and night irrigation flows. While field irrigation takes place only for 9 hours during day hours, water flows into the reservoirs during the remaining hours of the day. Field irrigations take place partly from the reservoirs and partly from the direct source at each scheme.

The water distribution systems at both schemes consist of a network of earthen (unlined) open canal system. Only 480 m length at the head of the main canal of Wonji-Shoa estate is concrete lined. For Metahara Scheme, the lined part of the main canal is about 400 m; of which 100 m lining is at the head of the main canal, while the remaining lining is at 3 and 5 km distance. At each division point there are flow control structure. The condition of these structures is much better for Metahara than Wonji-Shoa. While the flow control structures at Metahara Scheme are almost all functional, at Wonji-Shoa Scheme non-functional offtakes and check structures do exist. Absence of flow monitoring and recalibration of flow measuring structures has greatly affected irrigation water management at both schemes. Sedimentation and wear out are the major causes of mis-measurement of flows. Though poorly calibrated, at Metahara, the flow measurement facilities are in a better condition. At Wonji-Shoa, flow measurement is absent at the majority of offtakes. The flow control consists of underflow and overflow type of structures at both schemes. Water level regulators are commonly underflow type (sluice gates), while offtakes are Ronijn type overflow weirs. Flow regulation is manual and is labour intensive at each scheme. The design of the systems is in such a way that the tertiary flows are kept more or less constant for the whole irrigation period, and does not change with seasons. Tertiary offtake structures are generally operated twice a day; i.e. opening at 6:00 AM and closing at 3:00 PM. However, offtakes to secondary and branch canals are operated relatively more frequently to enable division of the flow into different parts of the systems as per the field water demands.

The design discharge of tertiary offtakes (sub-laterals) at Wonji-Shoa Scheme is 75 l/s, while that at Metahara Scheme is 200 l/s. Furrow is the method of field irrigation at both schemes with furrow inflow discharge of 5 l/s. The water distribution system is continuous in the main system and rotational at tertiary level. Hence, 15 and 40 furrows are irrigated at the same time at Wonji-Shoa and Metahara schemes respectively from the tertiary flow. An irrigation crew consisting of 3 persons is responsible for managing the tertiary flows both at Wonji-Shoa and Metahara schemes. Recently, a plastic pipe with openings (outlets) called hydro-flume has partially replaced field canals that feed furrows at Metahara Scheme (Figure 3.4). The hydro-flume has several advantages over a field canal. The main advantage is that it avoids run-off at the tail end of the canals and seepage in the canal. Secondly, it substantially reduces irrigation labour for operating siphons. Generally, it improves water management within a tertiary unit.

   a.   Siphons supplying water to furrows   b. Hydroflume supplying furrows
     Figure 3.4 Siphons versus hydroflume feeding furrows at Metahara Scheme

3.4.4   Water abstraction records (trends) in the large-scale schemes

Water abstraction (diversion or pumping) from Awash River is being measured by each scheme using a stage-discharge relation at a control section in the head reaches of the canals. Records of water stage (discharge) are being taken twice a day. At Wonji-Shoa Scheme, a calibrated staff gauge was fixed at the bank of the head reach, from which discharges are directly read. At Metahara Scheme, water levels are read from the staff gauge and are converted to discharges using an established Q-h relation. Discharges are being measured for two purposes: as means of matching field demands and supplies, and for billing purposes (payment of irrigation water fee to the Awash Basin Authority). Records of water abstraction of the two schemes are shown in Table 3.2.

3.4.5   Land and water productivity

Wonji-Shoa and Metahara schemes exclusively grow sugarcane for sugar production. The average cane yield at the schemes is 160 tonnes/ha (Behnke and Kerven, 2013). Land productivity from output calculated based on the net revenue from unprocessed cane is generally extremely low. However, processing cane and sugar production increases the net land productivity by about 550%. The existing water productivity calculated from the unprocessed cane based on net revenue is as low as 0.017 US\$/m^3 for Metahara scheme, while that calculated from sale of sugar based on net revenue is 0.091 US\$/m^3. Both land and water productivities at the sugar estate schemes is among the least for irrigated agriculture when valued based on local cane prices. Table 3.3 shows data on productivity of the schemes.

Table 3.2. Annual water abstraction records of the large-scale schemes

Year	2003	2004	2005	2006	2007	2008	2009	2010	Avg.
Wonji-Shoa, Mm^3	122	123	117	119	131	112	126	122	122
Metahara, Mm^3	186	192	191	194	178	199	187	189	190

Table 3.3. Productivity and output of the sugar estate schemes (cane and sugar)

Description	Wonji-Shoa	Metahara
Annual yield, tonnes/ha	165	161
Annual net revenue (cane), US$	2,350,000	3,140,000
Annual net revenue (sugar), US$	13,100,000	17,400,000
Output (cane), US$/ha	428	425
Output (sugar), US$/ha	2,390	2,350
Water use (cane), $US\$/m^3$	0.019	0.017
Water use (sugar), $US\$/m^3$	0.108	0.091

3.5   Community managed schemes

3.5.1   Brief history of Golgota and Wedecha irrigation schemes

Golgota and Wedecha community managed schemes are also located in Awash River Basin in the central part of Ethiopia (Figure 3.1). Geographically, Golgota is situated at 8°39'N and 39°45' E. It is found in Upper Awash Valley in between Wonji-Shoa and Metahara Schemes. This scheme was established during 1976 when the government of Ethiopia nationalized private farms for horticultural crops in the area (Girma and Awulachew, 2007). Later the scheme was handed over to the Ethiopian Red Cross Society. Since 1991 the scheme has been serving smallholder farmers. Currently, the irrigation scheme is being managed by the community (water users) themselves. The nominal command area of Golgota Scheme is 600 ha.

Wedecha Scheme is situated in the central highlands, about 50 km east of Addis Ababa. Geographically, it is located at 8°50'N and 38°57'E. It was established early 1980s by the then Ethiopian Water Resource Development Authority (EWRDA) in collaboration with Cuban Civil Mission. However, only the headwork was constructed and water distribution facilities were not completed. The remaining infrastructure (canals, flow control structures and farm outlet works) were later constructed and completed by the Ethiopian Water Works Construction Authority in 1984. Wedecha Scheme has two sub-systems, called Godino and Gohaworki located on the left and right banks. The combined nominal command area of the scheme is about 360 ha.

3.5.2   Physical features of the schemes

Climate

Climatic conditions have a vital impact on performance related factors, such as water and irrigation demand and water supply. Data of importance includes temperature, rainfall, wind speed, humidity and sunshine hours, all collected from a meteorological station nearby each scheme. The mean annual rainfalls are 583 mm and 820 mm respectively for Golgota and Wedecha schemes. Golgota Scheme is situated in the semi-arid Rift Valley region, while Wedecha Scheme is located on the Western edge of the Rift Valley lying in the temperate humid climate. A summary of the meteorological data (mean monthly values) at the two schemes are given in Table 3.4.

Table 3.4. Meteorological data at Golgota and Wedecha schemes

	Jan	Feb	Mar	Apr	May	Jun	Jul	Aug	Sep	Oct	Nov	Dec
Golgota												
Temp., °C	22	23	25	26	27	28	26	25	25	24	22	21
Rainfall, mm	35	12	57	41	26	32	139	140	46	39	5	11
Humidity, %	56	56	56	56	48	46	58	63	60	48	50	54
Wind speed, km/d	122	135	133	134	157	239	252	200	138	120	122	121
Sunshine, hrs	8.8	8.5	8.2	7.4	9.0	7.4	6.9	7.3	7.5	8.7	8.7	8.8
Wedecha												
Temp., °C	18	19	20	21	21	20	19	19	19	18	17	17
Rainfall, mm	10	28	49	57	51	90	211	197	93	21	10	3
Humidity, %	50	47	47	50	49	58	69	71	66	51	46	48
Wind speed, km/d	169	194	196	194	193	125	111	127	107	178	192	197
Sunshine, hrs	8.7	8.2	7.8	6.9	8.0	6.4	5.0	5.7	6.8	8.8	9.5	9.3

Beneficiaries (water users)

The beneficiaries of these schemes are smallholder farmers. Though the water users used to grow mainly food crops for their own subsistence, this trend has changed over the past decade. While the crop 'teff' (small Ethiopian grain) is a staple food crop at the schemes, it is rainfed. Irrigated crops at each scheme are both for subsistence and markets. These schemes have brought about major improvements in the rural livelihoods and living conditions of the beneficiaries compared to rainfed areas. The total number of beneficiary households of Golgota Scheme is 460 and that of Wedecha Scheme is 380.

Landholding and other characteristics

Landholding is one of the factors constraining agricultural output particularly in smallholder irrigation schemes and affects land and water productivity. While in some schemes water is a limiting factor, in others irrigable land becomes decisive. In Ethiopia, about 65% of farming households operate land sizes of less than 1 ha; while about 40% rely on a land size of 0.5 hectare or less (Gebreselassie, 2006). The average size of landholding of smallholder households in Ethiopia is 0.7 ha. In fact, landholding size is one of the major factors that constrain farm income and the level of household food security particularly in the Ethiopian highlands. For Golgota and Wedecha schemes, landholding sizes do not show significant variations in each scheme as all the users are smallholder. The average landholding for farmers of Golgota Scheme is 1.2 ha. The two sub-systems of Wedecha Scheme have different landholding sizes; with averages of 0.9 ha and 0.3 ha for Godino and Gohaworki respectively. The size of landholding at Wedecha Scheme is relatively small and it is one of the major limiting factors for increasing household income and improving livelihoods at the scheme. Some physical characteristics of each scheme are given in Table 3.5.

Table 3.5. Characteristics of the community managed schemes

Characteristics	Golgota	Wedecha
Source of water	River Awash	Wedecha stream
Means of water abstraction	River diversion	Small earth dam
Nominal irrigable area, ha	600	360
Number of beneficiary households	460	380
Major crops	Onion, Tomato, Maize, Cabbage	Onion, Tomato, Maize, Lentil, Sugarcane

Roads and access to market

Both the schemes are connected to nearby market centres via dry weather road. The nearest major market centre for Golgota scheme is Adama, which is at 85 km away. However, there are small rural market centres in the vicinity. For Wedecha scheme, the nearest major market is Debre Zeit at 15 km distance. Access from the schemes to the major market centres during rainy seasons is one of the major challenges of farmers. As such, farmers are in some cases obliged to sell their produce for incompetent prices at local small markets due to lack of easy access.

3.5.3 Source and availability of irrigation water

Golgota Scheme is supplied with water from Awash River with temporary diversions made of stones, sand bags and wire mesh for raising the stage in the river (Figure 3.5). With the absence of a permanent diversion structure, the temporary structure is being frequently washed away by floods. Still, farmers face little difficulty to divert water as the river flow is large compared with the irrigation water to be diverted. For Wedecha Scheme, the source of irrigation water is Wedecha Reservoir, with live storage capacity of 14.2 Mm^3 at construction. The reservoir was created by an earthen dam and supplies water through a piped outlet under the dam (Figure 3.6). In addition to Godino and Gohaworki sub-systems, Wedecha Reservoir also supplies water to other small schemes in the vicinity. Capacity of the reservoir is sufficient to supply water to these two schemes and other local schemes in the district in a simple supply-demand assessment. However, inadequate irrigation water management, poor irrigation service, low water productivity and poor management of irrigation infrastructure are the major problems.

a. Temporary diversion for Golgota Scheme   b. Regulating gates on main canal banks
Figure 3.5. Water acquisition and control structures for Golgota Scheme

a.   Wedecha Reservoir outlet   b. Diversion structure to Godino sub-system
Figure 3.6. Water acquisition and control structures for Wedecha Scheme

3.5.4   Water management, flow control and field irrigation practices

For Golgota Scheme, at 500 m from the head of the main canal (temporary diversion), there are sluice gates on the bank of the canal to regulate the flow (Figure 3.5). These sluices are used to release excess water from the canal back to the river and to scour sediment entering at the head of the canal. Water is conveyed through the earthen main canal and is distributed through three main tertiary offtakes equipped with sluice gates. Irrigation water management is a sole responsibility of the water users themselves at Golgota Scheme. Farmers are responsible for water diversion, distribution and scheduling at field levels. Though border flooding is also being practiced occasionally, furrow is the most common method of field irrigation at Golgota Scheme.

Water is conveyed from the reservoir supplying Wedecha Scheme through the natural channel of the river for about 5 km. From there, a diversion weir with offtake diverts water from the channel to the right bank that supplies water to Gohaworki Sub-system. At 1 km downstream of the first weir is the second diversion weir with offtakes on the left bank supplying water to Godino Sub-system. The regulating gates at both of these offtakes were demolished by farmers. Currently, flow into the canals is regulated at the offtakes with stones and wooden logs. Water is diverted into rectangular masonry lined canals at both offtakes and is distributed using poorly constructed and maintained earthen canals. While release of water from the reservoir is decided by a regional governmental irrigation agency, water users are responsible for distribution and irrigation scheduling at field levels. Like at Golgota scheme, furrow is the dominant method of field irrigation at both sub-systems of Wedecha scheme.

3.5.5   Existing institutional setups for water management in the schemes

Irrigation water management institutions in the community managed schemes have existed since their operation. However, these served only scheduling water deliveries at field levels and resolving conflicts on water sharing. During the last 10 years, water management at both schemes has improved. However, issues such as service satisfaction, equity, water productivity and schemes sustainability still need special attention. At Golgota Scheme, the water users association (WUA) is responsible for all aspects of irrigation water management, including diversion from the source, field delivery scheduling and routine operation and maintenance. Irrigation water fee is collected by the WUA for routine and some major maintenance activities. However, there is no fee paid to an agency for irrigation water. Water users in this scheme have never paid irrigation water fees to an external irrigation agency except that required for running the scheme themselves. The WUA has an executive committee comprising of Chairperson, Vice chair, Secretary, Treasurer, and five members.

Unlike at Golgota Scheme, at Wedecha Scheme a regional government agency (Oromiya Water Works Construction Enterprise) and water users associations are involved in irrigation water management. It is practically dual-managed. However, it is still termed community managed scheme, because the agency is only responsible for making decisions on the release of water from the reservoir. Of course, decisions are based on requests from water users and water availability. A preset water release schedule is prepared by the agency on a monthly basis, which is subject to modification based on a request from the WUA. The agency is responsible for the proper functioning of the dam and reservoir and its outlet works. Maintenance related to the headwork is also made by the agency. Previously water users used to pay a fee only for the routine maintenance activities through water users associations. However, since 2010, an annual fee of ETB 862 (US$ 48) per hectare was introduced by the agency for the services it renders related to irrigation water management. At each sub-systems of Wedecha Scheme exists a separate water users association. The WUAs are responsible for field water delivery schedules, collection of fees, scheduling routine maintenance, resolution of conflicts on water sharing between farmers or groups of farmers, penalty procedures for failure to abide by the rules and regulations, etc.

4. COMPARATIVE INTERNAL AND HYDRAULIC PERFORMANCE EVALUATION OF IRRIGATION SCHEMES

4.1 Introduction

Performance assessment has been an integral part of irrigation since man first started harnessing water to improve crop production (Bos et al., 1994). Performance evaluation of irrigation schemes has specially been an important and active field of research during the last few decades. Several approaches and methodologies have been developed for assessing irrigation performance from different perspectives. With limited water and land resources availability for the required global increase in food production, improving the productivity of existing irrigation schemes has got an increasing attention. Global cereal production has to duplicate in the next 25 to 30 years, while 80-90% of this increase would have to be realized from the existing agricultural land (Schultz, 2012). This would be through increasing land and water productivity by various interventions as intensifying irrigation, employing water saving mechanisms, innovating irrigation management, improved drainage, etc. All these have to improve the performance of the existing irrigation and drainage schemes in order for these schemes to achieve the objectives. Performance evaluation in irrigation is a systematic observation, documentation and interpretation of the management of an irrigation scheme, with the objective of ensuring that the input of resources, operational schedules, intended outputs and required actions proceed as planned (Bos et al., 2005).

Performance of irrigation schemes is assessed for a variety of reasons. It can be to improve scheme operations, to assess progress against strategic goals, as integral part of performance-oriented management, to assess the health of a scheme, to evaluate impacts of interventions, to better understand determinants of performance, to diagnose constraints and to compare the performance of a scheme with others or with the same scheme over time (Molden *et al.*, 1998). Recently, benchmarking in the irrigation and drainage sector was identified as a useful tool for continuous improvement. Benchmarking in irrigation schemes implies improving all aspects of service delivery and resource utilization by comparison with other schemes (Malano et al., 2004). However, benchmarking goes beyond comparison and involves several steps in the change process.

4.2 Framework for performance assessment

Irrigation system performance assessment needs a framework to adequately guide the work and for the stakeholders to effectively use the outcomes from performance assessment. The purpose of the framework is to form a link between repeated actions in such a way as to provide a learning experience for the manager that allows things to be done better in each successive iteration (Bos et al., 1994). The framework defines why the performance assessment is needed, what data are required, what methods of analysis will be used, who is the performance assessment for, etc. (Bos et al., 2004). Without a suitable framework the performance assessment programme may fail to collect the necessary data, and may not provide the required information and understanding to the user. Performance assessment is based on collection, analysis and interpretation of data related to irrigation management and irrigation service delivery.

Performance assessment of irrigation management can be viewed from two perspectives: operational performance and strategic performance. Operational performance evaluation help irrigation management to address the question 'Am I doing things right?'; while strategic performance assessment addresses the question 'Am I doing the right thing'?. However, before one can start with the actual data collection, at least the following issues would have to be addressed:

- the need for assessment;
- the overall objective;
- specific objectives;
- whom is the assessment for.

The need outlines the rationale of the performance assessment or its purpose. The need in general would have to be addressed in a wider context instead of a specific and a single purpose. Performance evaluation needs to achieve an objective in a broader sense, which might be of interest to one or more stakeholders in the sector. Hence, next to the need, the general objective has to be clearly defined. In order to achieve the overall objective, it has to be split into specific objectives. Specific objectives enable one to deal with smaller problems, whose cumulative would make up the bigger picture of the problem. The information derived from irrigation performance assessment could be of primary importance to one or more stakeholders in the sector; for instance researchers (scientific community), government, the scheme management, water user associations, farmers, funding agency, etc. The actual evaluation of the performance could, however, be done by a university, research institute, government agency, etc.

4.3   Internal and external performance indicators

Diagnosis of irrigation performance essentially has to incorporate all aspects of the irrigated agricultural system including institutional setups, resources used, services delivered and agricultural outputs. Performance indicators can be broadly categorized into internal and external indicators to describe the above mentioned aspects. Internal indicators are used to assess the performance of the internal processes and irrigation services. They are concerned with operational procedures of the systems, institutional setups for management, irrigation infrastructure and water delivery services. Internal indicators enable comprehensive understanding of the processes that influence water delivery service and the overall performance of a system (Renault et al., 2007). Hence, they are useful to show what would have to be done to improve the internal and hence the external performance. External indicators on the other hand evaluate inputs and outputs to and from irrigation schemes. They are generally meant to evaluate the efficiency of resource use (land, water, finance) in irrigated agriculture. External indicators can be best used as part of a strategic performance assessment and benchmarking performance of schemes (Burt and Styles, 2004).

4.4   Comparative irrigation performance evaluation

Comparative performance indicators enable to see how well irrigated agriculture is performing at different scales, i.e. at the scheme, basin, national or international scales. Comparative performance has a set of advantages for stakeholders in the irrigation and drainage sector, including policy makers, irrigation managers, researchers, farmers and donors.

4.4.1 Rationale for comparative performance assessment

Land and water, the two principal resources for irrigated agriculture are limited, and in some countries, critical. Irrigated agricultural production needs to improve the utilization of these increasingly scarce resources. Comparative (external) performance evaluation enables irrigation stakeholders to see how productively land and water resources are used for agriculture. Cross comparison of schemes assists in answering crucial questions related to irrigated agriculture. Some of these questions are: which types of irrigation systems enable higher benefits from limited water and land resources? Where, how and how much investment needs to be made in irrigation? Which water acquisition and decision making best meets and responds to heterogeneous irrigation schemes with cropping freedom? Moreover, it provides a systematic means of tracking performance in individual irrigation schemes.

4.4.2 Comparative performance and indicators

Comparative performance assessment in irrigation schemes is possible through use of comparative indicators. External indicators are those indicators based on outputs and inputs from and to an irrigated agricultural system (Molden et al., 1998). However, they practically provide little or no detail on internal processes that lead to the output. Unlike internal indicators, which relate performance to internal management targets, external indicators enable comparison between different regions, different infrastructure and management types, and different environments. Moreover, the trend in performance of a specific scheme can be compared over time. Internal irrigation performance is also linked to farmers' level of satisfaction by some authors (Ghosh et al., 2005; Kuscu et al., 2008). Comparative performance is of primary importance to policy makers and managers making long-term and strategic decisions, and researchers looking for relative differences between irrigation systems. Although, in its very concept, external indicators link outputs to inputs, there are indicators for comparative purposes that are not necessarily based on outputs and inputs. Examples are water supply, financial and physical sustainability indicators. The International Water Management Institute (IWMI) proposed a minimum number of comparative indicators (Molden et al., 1998): 4 for agricultural output, 2 for water supply, 1 for delivery capacity and 2 for financial.

In this research, three groups of relevant comparative performance indicators were used to compare the performance of the two community managed irrigation schemes: two from IWMI, to which a third group called 'physical sustainability' indicators, is added.

4.4.3 Water supply indicators

The water supply indicators are based on irrigation and water supply/delivery measurements being related to water demands or irrigated area. The three indicators that will be considered under this group are:

Annual irrigation water delivery per unit irrigated cropped area (m³/ha)

The annual irrigation water delivery quantifies the volume of irrigation water actually delivered per unit area irrigated (Malano and Burton, 2001). In this study, delivered irrigation water at the command head will be considered instead of the diverted supply.

The cropped area will be the irrigated area of the schemes. It is given by:

$$AIDUIA = \frac{\text{Annual water delivered}}{\text{Irrigated cropped area}} \; (m^3/ha)$$

(4.1)

Where, *AIDUIA* is annual irrigation water delivery per unit irrigated cropped area.

Annual relative water supply

The annual relative water supply is the ratio of total annual water supplied (irrigation plus rainfall) to the annual crop water demand. It signifies whether the water supply is in short or in excess of demand:

$$ARWS = \frac{\text{Annual water supply}}{\text{Annual crop water demand}} \; (m^3/m^3)$$

(4.2)

Where, *ARWS* is annual relative water supply.

Annual relative irrigation supply

The annual relative irrigation supply is the ratio of annual irrigation supply to annual irrigation demand. Irrigation water is a scarce resource in many irrigation schemes and may be a major constraint for production. This indicator is useful to assess the degree of irrigation water stress/abundance in relation to irrigation demand. It is given by:

$$ARIS = \frac{\text{Annual irrigtion supply}}{\text{Annual irrigation demand}} \; (m^3/m^3)$$

(4.3)

Where *ARIS* is annual relative irrigation supply.

4.4.4   Agricultural output indicators

Agricultural output indicators can be subdivided into land productivity and water productivity indicators. Six relevant indicators, two for land productivity and four for water productivity were considered under this group of indicators for this study. The outputs of agricultural production in this paper were based on local prices.

Output per unit irrigated cropped (harvested) area (US$/ha)

The output per unit irrigated cropped area (output per unit harvested area) quantifies the total value of agricultural production per unit of area harvested during the period of analysis. The annual harvested area depends on the intensity of cropping (irrigation intensity). The area is the sum of all the areas under crops during the year in this case. This indicator is not affected by the intensity of cropping (irrigation). However, it can also indirectly indicate the degree of irrigation water availability. In addition to water availability, soil type and fertility, land suitability, crop variety and agricultural inputs do have significant impact on output and hence on land productivity. It is given as (Molden et al., 1998; Malano et al., 2004):

$$OPUIA = \frac{\text{Value of annual production}}{\text{Harvested area}} \quad (US\$/ha)$$
(4.4)

Where, $OPUIA$ is output per unit irrigated cropped (harvested) area.

Output per unit command area (US$/ha)

The output per unit command area is the value of agricultural production per unit of nominal area which can be irrigated. Smaller values of this indicator can also imply, although not necessarily, less intensive irrigation and vice versa. It is particularly important where land is a constraining resource for production (Molden et al., 1998). It is given as:

$$OPUCA = \frac{\text{Value of annual production}}{\text{Nominal area}} \quad (US\$/ha)$$
(4.5)

Where, $OPUCA$ is output per unit command area.

Output per unit irrigation water supply (US$/m³)

The output per unit irrigation water supply tells on how well the total annual diverted irrigation water from a source is productive. Irrigation water supply includes conveyance (seepage) losses in canals, and hence it is generally measured at the intake from the source or at diversion. In areas where water is scarce, water management aims to increase the output per drop of irrigation water:

$$OPUIS = \frac{\text{Value of annual production}}{\text{Diverted annual irrigation water}} \quad (US\$/m^3)$$
(4.6)

Where, $OPUIS$ is output per unit irrigation water supply or diverted.

Output per unit irrigation water delivered (US$/m³)

The output per unit irrigation water delivered is meant for the value of production per unit volume of annual irrigation water delivered to the head of command area. It is different from irrigation supply as it does not include losses in conveyance systems. It is a useful comparative indicator because it addresses output per drop of irrigation water actually delivered to the user. Inefficient water use results in lower values of this indicator:

$$OPUID = \frac{\text{Annual value of production}}{\text{Delivered annual irrigation water}} \quad (US\$/m^3)$$
(4.7)

Where, $OPUID$ is output per unit irrigation water delivered.

Output per unit water supply (US$/m³)

The output per unit water supply is for the output per unit of total annual volume of water (effective rainfall + irrigation) diverted to the system. It gives a sound comparison between irrigation schemes with different rainfalls, because gross water supply was considered:

$$OPUWS = \frac{\text{Annual value of production}}{\text{Total water supply}} \quad (\text{US\$}/\text{m}^3) \qquad (4.8)$$

Where, $OPUWS$ is output per unit water supply/diverted.

Output per unit water consumed (US\$/m³)

The output per unit water consumed informs on the output per unit annual volume of water consumed by actual evapotranspiration (ET). Its value is highly dependent on climate. Moreover, less consumptive use coefficient due to water losses does not affect its value; as only the water consumptively used by the crops is considered. It is given as (Molden et al., 1998):

$$OPUWC = \frac{\text{Annual value of production}}{\text{Water consumed by ET}} \quad (\text{US\$}/\text{m}^3) \qquad (4.9)$$

Where, $OPUWC$ is output per unit water consumed.

4.4.5 Physical sustainability indicators

Two indicators are of relevance under the group of physical sustainability indicators as was enumerated by Yercan et al. (2004) and Şener et al. (2007).

Irrigation ratio

Irrigation ratio is the ratio of currently irrigated area to irrigable command (nominal) area. It tells the degree of utilization of the available command area for irrigated agriculture at a particular time. Shortage of irrigation water, lack of irrigation infrastructure, lack of interest on irrigation due to less return, reduced productivity due to problems such as salinization/waterlogging, etc, could result in under utilization of land. On the other hand, cropping intensity, a ratio of annual cropped area to nominal area is indicative of annual land utilization. Burton et al. (2000) state that cropping intensities from 100 to 200% are considered good; whereas an inferior figure is low. Irrigation ratio is expressed as:

$$Irrigation\ ratio = \frac{\text{Irrigated area}}{\text{Command (nominal) area}} \quad (\text{ha}/\text{ha}) \qquad (4.10)$$

Sustainability of irrigated area

Sustainability of irrigated area is the ratio of currently irrigated area to initially irrigated area when designed (Bos, 1997). It is a useful indicator for assessing the sustainability of irrigated agriculture. Lower values of this indicator would mean abandonment of lands which were initially irrigated; and hence, indicate contraction of irrigated area over time. On the other hand, values higher than unity indicate expansion of irrigated area and would imply more sustainable irrigation:

$$Sustainability\ of\ irrigated\ area = \frac{\text{Currently irrigated area}}{\text{Initially irrigated area}} \quad (\text{ha}/\text{ha}) \qquad (4.11)$$

4.5   Internal irrigation service performance evaluation

Internal (process) irrigation performance assessment refers to the day-to-day operational process of schemes. Internal process indicators are evaluated relative to the targets set based on specific objectives of the systems management. In many cases, these indicators are specific to individual systems, and different indicators may apply to systems with different internal processes. Internal (process) indicators help the system manager to answer the question 'Am I doing things right?' (Murray-Rust and Snellen, 1993).

4.5.1   Irrigation as a service

In a business, when a service, service provider and service receiver do exist, there happens a flow of service from the provider and a payment back for the services. Irrigation is regarded as a service provided to farmers (water users) from irrigation agencies, either governmental or private. In irrigation systems with s service provision setup, the management objectives would have to address the agreed upon services. Service Oriented Management is a managerial approach that focuses on the supervision and control of the delivery of a service from a service provider to a service receiver (Renault, 2008). The main purpose of internal (process) performance assessment is to improve the irrigation service to users. These indicators are of significance to irrigation service providers and managers as means of monitoring their services and management decisions.

4.5.2   Water users' qualitative view of irrigation service

Evaluation of internal process indictors is often not easy due to their data intensive nature. Conventional evaluation needs routine quantitative data on water deliveries, timing and duration, whose observation and measurement is given less priority, particularly in least developed countries. The water users' view could in this case be used to assess the quality of irrigation service. Qualitative data related to the service are relatively easier and cheaper to collect from water users, and best be used to reflect on the internal processes in the absence of quantitative data on water deliveries. This gives primary attention to the water users who are the direct stakeholders in the irrigation system.

4.5.3   Irrigation service utility

Utility of an irrigation service refers to the quality of the services with respect to water users. Utility is comprised of different components related to irrigation water deliveries. It can be described by reliability, timing, flow rate, duration of supply and tractability (El-Awad, 1991). In systems where water delivery measurements are available, utility can be measured with respect to a standard criterion of performance. However, in small community managed schemes, like in Golgota and Wedecha, where data on irrigation water delivery and timing are not available, qualitative data could be made use of. However, utility in many cases needs to be expressed in quantitative terms to make clear distinction. For this, a methodology based on fuzzy set theory was used in this thesis.

4.5.4   Fuzzy set theory and its application for irrigation service utility evaluation

There is large uncertainty (fuzziness) associated with irrigation service utility from qualitative data on farmers' perceptions. Unlike the classical set theory, in a fuzzy set, partial membership of an element of a set is possible. In a classical theory of a set the degree of membership of an element in a set is either 0 or 1, which means an element either belongs or does not belong to the set. In a fuzzy set, the degree of membership of an element is designated by its support (membership function) ranging over the real range from 0 to 1. In this theory, fuzzy linguistic expressions and support functions are designated for each variable. The fuzzy set theory is useful to deal with real situations in a wide range of domains in which information is incomplete, imprecise or associated with fuzziness.

For a set X of objects denoted generically by x, a fuzzy set A in X is defined as:

$$A=\{x, \mu(x)\} \tag{4.12}$$

Where, $\mu(x)$ is a membership function. For instance, the fuzzy set A for x=1 to 5 is denoted as:

$$A=\{1|\mu(1), 2|\mu(2), 3|\mu(3), 4|\mu(4), 5|\mu(5)\} \tag{4.13}$$

Where, $\mu(1)$ to $\mu(5)$ are support functions of the respective variables of the fuzzy set. One may select different support functions to represent a particular expression with fuzziness. However, the choice of the supports needs to be rational. Ghosh et al. (2005) state that the choice of a support to represent a fuzzy linguistic expression may depend on the particular problem being taken into consideration. Sam-Amoah and Gowing (2001) and Ghosh et al. (2005) applied a universe of U from one to five with different supports for linguistic expressions from very good/very high to very bad/very low.

In irrigation service evaluation the fuzzy set concept can be used to aggregate the opinions (linguistic expressions) of all sampled farmers regarding the utility factors and sub-factors. Sam-Amoah and Gowing (2001) applied this concept for analysing the performance of irrigation systems from qualitative data (responses of water users). Ghosh et al. (2005) also applied a methodology of irrigation service utility assessment from farmers' perspective to evaluate the performance of irrigation at Orissa, India.

Applications of this theory include artificial intelligence, computer science, medicine, control engineering, decision theory, expert systems, management science, operations research, etc. (Zimmermann, 2010). Moon et al. (2007), for instance applied fuzzy sets for a performance appraisal and promotion ranking for a case in military organizations. In their study, they proposed seven linguistic variables with which the candidates were scored. The expressions they used were very good, good, medium good, medium, medium bad, bad, very bad. They considered fuzzy numbers from zero to ten as members of the fuzzy set with support functions as shown in Figure 4.1.

Opinions of water users regarding the irrigation services they receive will be different. Hence, their opinions need to be aggregated over the area extent where the service levels are required. There are different methods of aggregation; however the one used by Ghosh et al. (2005) is as follows. According to this method of aggregation, the average support is calculated for each element of the set; it is added to the maximum support in the set and divided by 2. This gives the aggregated support of each element of the fuzzy set.

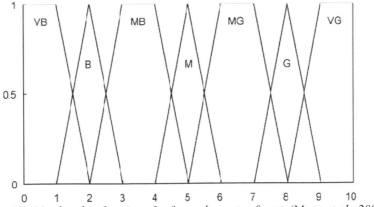

Figure 4.1. Membership functions for fuzzy elements of a set (Moon et al., 2007)

The fuzzy set, which is formed with the aggregated support as discussed, may not in many cases have exactly the supports which fit to one of the fuzzy linguistic expressions. So, the obtained fuzzy set can be approximated to the nearest expression. If A_j represents the fuzzy linguistic expressions with different supports and B is the fuzzy set obtained by aggregation of the farmers' opinions, the difference D is calculated, and then the expression with the least difference is the one which approximates the set B (Sam-Amoah and Gowing, 2001; Ghosh et al., 2005)

$$D(B, A_i) = [\sum\{\mu B_i - \mu A_{j,i}\}^2]^{1/2} \tag{4.14}$$

Where, μ is the support of the element i of a set. Once the water users' opinions on irrigation services were aggregated, it tells the suitability of the service with linguistic expressions. However, this is not suitable for using in decision making related to improvement of system management. It would be more convenient if the qualitative fuzzy expression is converted to a numerical scale. El-Awad (1991) calls this numerical value as farmer utility (*FU*) and ranges from 0 to 1. Utilizing the same method, the *FU* for a universe consisting of N elements shall be evaluated from:

$$FU = \left\{1/(N-1)\right\} * \left[\frac{\{\sum(i-1)*\mu_i\}}{\sum \mu_i}\right] \tag{4.15}$$

Where, *FU* is farmer utility, i is possible values of the universe U and μ_i is the support of the element i of the fuzzy set.

4.6   Hydraulic/water delivery performance in irrigation schemes

Irrigation systems are ideally designed to meet field irrigation requirements uniformly in full, without wasting water. However, due to technical and operational reasons, irrigation systems often do not meet the design objectives. Hydraulic performance refers to adequacy of conveyance, distribution and delivery of irrigation water in spatial and temporal scales. Hydraulic performance is generally measured against some criteria, for which indicators can be developed. Factors such as adequacy, operational efficiency, equity, reliability, timeliness, delivery performance ratio (DPR), etc. have been proposed and used by a large number of researchers, including Molden et al. (1998); Unal et al. (2004); Tariq et al. (2004); Vos (2005); and others. In many large-scale manually operated gravity irrigation schemes in least developed and emerging countries, poor hydraulic performance remains a major challenge. This is largely associated with lack of knowledge and information regarding the complex hydraulic behaviour of the systems. Effective canal operation for adequate hydraulic performance requires knowledge of when, where and how operations should be made, and understanding the effects of operational decisions (Renault, 2000).

Inadequate operation of large-scale gravity irrigation schemes have highly affected the benefits expected from these schemes. The first and most apparent consequence of ineffective operation is inequity in water delivery at head, middle, and tail reaches of the system. This results in over-supply in some parts of the scheme and under-supply in other parts, which brings about differences in irrigation services among different groups of water users. The second consequence of ineffective operation is loss of significant quantity of water either as over application or run off losses at tail end, which results in low water use efficiencies. This in turn results in low water productivity and puts limitations on the water resources available to irrigate more lands to meet rising food demands. The third consequence of ineffective operation is related to non-sustainability of the schemes due to waterlogging and salinization. This is a typical problem particularly in large-scale schemes in arid and semi-arid regions with poorly drained soils.

Operation of canal irrigation systems involves decision making on operation of flow control and delivery structures (offtakes and water level regulators) and anticipation of the resulting hydraulic responses (outputs). Operation is an external perturbation (input), while results, like change in discharge or water levels, are outputs. Thorough knowledge of the relationship between inputs and outputs is basic for effectiveness of operation. Hydraulic principles govern the way each component of the system is being affected by a perturbation. However, canal systems are complex, and their interactive operation and hydraulic behaviour are often less understood by operators. Thus, there is a need for relatively easier means to aid effective canal operation.

Hydraulic simulation during the last two decades, has contributed a lot to the evaluation of complex hydraulic performances and to decision making in irrigation system operation. Hydraulic modelling has been used and verified to be of useful tools to assist operational decision making in large-scale irrigation systems by researchers Kumar et al. (2002), Shahrokhnia and Javan (2005), Tariq and Latif (2010), etc. Some of the useful hydraulic models developed and have been in use during the past two decades for simulation of flows in irrigation networks include CANALMAN, SIC, CARIMA, MODIS, USM, DUFLOW, HEC-RAS, etc. In this study, DUFLOW, will be applied as a tool for simulating the flow for different operational setups.

DUFLOW is a one-dimensional unsteady hydrodynamic model for simulating flows in natural and artificial surface water systems (STOWA, 2004). The model was jointly developed by the Rijkswaterstaat, Wageningen University, Delft University of Technology and UNESCO-IHE. It is designed to cover a large range of applications, such as propagation of tidal waves in estuaries, flood waves in rivers, operation of irrigation and drainage systems, etc. Basically, the model simulates free flow in open channel systems, where control structures like weirs, pumps, culverts and siphons can be included. It is based on the one-dimensional partial differential equations that describe non-stationary flow in open channels (Abbott, 1979). These equations (Saint Venant equations) are for conservation of mass and momentum and read as follows:

$$\frac{\partial Q}{\partial x}+\frac{\partial A}{\partial t}=0 \tag{4.16}$$

$$\frac{1}{A}\frac{\partial Q}{\partial t}+\frac{1}{A}\frac{\partial}{\partial x}\left(\frac{Q^2}{A}\right)+g\frac{\partial y}{\partial x}-g(S_o\text{-}S_f)=0 \tag{4.17}$$

Where Q is discharge (m^3/s), A is cross sectional area (m^2), y is flow depth (m), x is distance in the flow direction (m), t is time (s), v is flow velocity (m/s), g is acceleration due to gravity (m/s^2), S_f is friction (energy) slope (m/m) and S_o is bed slope (m/m).

The equation for conservation of mass states that if there is a change of water level, this will be a result of local inflow minus outflow in that reach. The momentum equation on the other hand states that the net change of momentum is the result of interior and exterior forces such as friction and gravity. Duflow solves these equations for conservation of mass and momentum in space and time. The partial differential equations written as a system of algebraic equations using four-point implicit Preissmann scheme were solved by the model (STOWA, 2004).

Hydraulic performance could also be assessed based on routinely collected flow data. This requires monitoring of spatial and temporal routine flow data at various offtake and division points. Hydraulic performance evaluation from routine flow data can provide useful information regarding the existing level of performance; however, it does not provide options for quickly assessing alternative operations and their impacts. In this thesis, hydraulic performance of the two large-scale schemes will be evaluated in chapters 5, 6 and 7. While performance based on routinely measured flow data has been assessed for both schemes, hydraulic simulation was employed at Metahara Scheme only. Adequacy (relative delivery), efficiency, equity and reliability were used as criteria for hydraulic (water delivery) performance evaluation and corresponding indicators were employed. Table 4.1 shows the ranges of performance levels for these four water delivery performance indicators.

Table 4.1. Performance standards for water delivery performance indicators
(Molden and Gates, 1990)

Indicator	Poor	Fair	Good
P_A	< 0.80	0.80 – 0.89	≥ 0.90
P_F	< 0.70	0.70 – 0.84	≥ 0.85
P_E	> 0.25	0.11 – 0.25	≤ 0.10
P_D	> 0.20	0.11 – 0.20	≤ 0.10

5. HYDRAULIC PERFORMANCE EVALUATION OF WONJI-SHOA LARGE-SCALE IRRIGATION SCHEME

5.1 Water acquisition, distribution and delivery

Irrigation water for Wonji-Shoa Sugar Estate scheme is supplied via a pumping station on Awash River. Water is pumped 24 hours a day, delivered to a settling basin, from which it is conveyed and distributed by gravity. This scheme is located in the Upper Awash River Basin, where there are little irrigation developments upstream. Water is distributed to three major operational units (sections) of the scheme by three branch (secondary) canals known as Right, Left and Eastern branches (Figure 5.1). Five major night storage reservoirs balance the day and night hour pumping and field irrigation flows. Water is delivered to tertiary units mainly via overflow weirs (offtakes) and few underflow gates. In this thesis, the Right branch canal with a length of 9 km and irrigating an area of 2,000 ha was considered for assessment of the water delivery performance.

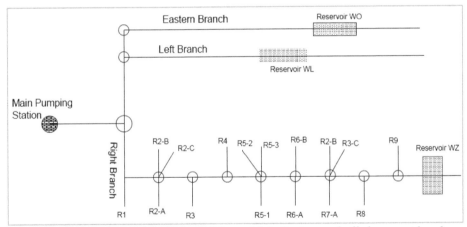

Figure 5.1. Layout of Wonji-Shoa Scheme Main System and offtakes considered

5.2 Methodology

5.2.1 Field data collection

Data collection basically included climate data, soil data, crop data and flow measurements. Climate data was required for determination of the scheme water and irrigation requirement, and was collected from the estate's research centre. Soil and crop data were collected from the plantation department of the estate, and were used for water requirement and irrigation scheduling. Flows were monitored on a daily basis at 16 offtakes along one of the secondary canals (Right branch) for three months. Of these, five offtakes are situated in the head, six in the middle and the remaining five in the tail reaches of the canal. The flows were measured in two consecutive years (2012 and

2013) for three months (January-March).

5.2.2 Flow measurement (current metering)

For offtake flow measurement, current meters were used. Flows were measured three times a day at each offtake in order to determine the average daily discharges. The discharges of the canals were computed using the Mean section method. In this method, the cross section of a canal is divided into a number of verticals, at which water depths and depth-averaged velocities are measured. The flow (q) between any two adjacent verticals is a product of the width between verticals (W), the mean of water depths of two adjacent verticals and the mean of the average velocities over those two verticals (Figure 5.2). The total discharge (Q) at the section is determined as a sum of the discharges in each sub section.

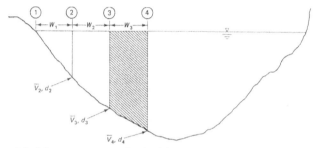

Figure 5.2. Mean section method of flow determination in open channels

$$Q = \sum_{x=1}^{n} W_x \left(\frac{\bar{V}_x + \bar{V}_{x+1}}{2} \right) \left(\frac{y_x + y_{x+1}}{2} \right) \qquad (5.1)$$

Where Q is total discharge, W is width between two adjacent verticals, $\tilde{V}$ is depth-averaged velocity, y is flow depth, and x is the number of verticals.

5.2.3 Indicators of water delivery performance

Four water delivery performance indicators, namely adequacy, efficiency, equity and reliability were used. These indicators could be evaluated for a single offtake, a group of offtakes in a sub-system, or for a whole irrigation scheme. Adequacy is an indicator for a water delivery system whether it attained a target or required water delivery over a certain period of time. The time frame to be considered could be daily, per irrigation turn, monthly, seasonally or annually as required. In this case, a period of three months is considered. It is given by:

$$P_A = \frac{1}{T} \sum_T \left(\frac{1}{R} \sum_R p_A \right) \qquad (5.2)$$

Where P_A is adequacy indicator aggregated over a region R and time T, p_A is a ratio of delivered to required flows at a point (offtake).

However, in the conventional adequacy indictor (Molden and Gates, 1990) and (Bos et al., 1994), it could be observed that when the delivered amount is in excess of the required, the adequacy indicator (Sakthivadivel *et al.*) takes a maximum value of 1.0. Hence, this indicator cannot capture performance in cases of excess water delivery as in the case of Wonji-Shoa Scheme. If not well addressed, in addition to the apparent inefficiency in water use, excess delivery can result in serious environmental concerns like waterlogging. (Kazbekov et al., 2009) reported adequacy indicator (Sakthivadivel *et al.*) with values higher than 1.0 in case of excess deliveries, for instance. With average groundwater levels of about 1 m below surface at Wonji-Shoa Scheme, excess water delivery and hence waterlogging has been a major problem. Hence, an indicator named 'relative delivery indicator', which could have values higher than 1.0 is used in this study, instead of adequacy indicator.

Efficiency is an indicator for water conservation property of an offtake or a system. It is actually an inverse (opposite) of adequacy (relative delivery). Equity implies the fairness of water delivery to different parts of an irrigation system. It indicates the spatial distribution of a water delivery. In large-scale gravity irrigation systems with manual operation of structures, ensuring equity is a major challenge for operators. The coefficient of variation (CV) of the ratio of delivered (Q_D) to required (Q_R) delivery over an area R and time T as proposed by Molden and Gates (1990) was used. Reliability is a term used for the capacity of a water delivery system to meet prior expectations. Reliability actually encompasses two elements: reliability of the delivery amount and reliability of the timing. Reliability (dependability) indicator as suggested by Molden and Gates (1990) is the coefficient of variation of delivered to required delivery over time T and region R. The indicator includes both predictability and variability of supply.

$$P_F = \frac{1}{T}\Sigma_T \left(\frac{1}{R}\Sigma_R p_F\right) \tag{5.3}$$

$$P_E = \frac{1}{T}\Sigma_T CV_R \left(\frac{Q_D}{Q_R}\right) \tag{5.4}$$

$$P_D = \frac{1}{R}\Sigma_R CV_T \left(\frac{Q_D}{Q_R}\right) \tag{5.5}$$

Where P_F, P_E and P_D are efficiency, equity and dependability indicators respectively; pF is a ratio of required to delivered flows at an offtake, T is time, R is region, CV_T and CV_R are coefficients of variation over time T and region R respectively.

5.2.4 Irrigation supply and demand determination

Daily measurements of irrigation supply to the scheme is being made by the estate using a graduated staff gauge *(Q-h)* fixed at the bank, in the head reach of the main canal. Monthly and hence annual irrigation supplies to the scheme were determined as a sum of daily supplies for 7 consecutive years (Table 5.1), for which the average annual supply was 121 Mm³. Irrigation demand was determined with FAO CROPWAT software, Swennenhuis (2006), using climatic and crop data. To do so, planting of sugarcane was distributed over 4 months from January to April, over one quarter of the total irrigated land for each. Total monthly and annual irrigation water demand was determined as a sum of the water requirements for each planting block. Monthly and annual irrigation demand for Wonji-Shoa Scheme is shown in Table 5.2.

Hydraulic and operational performance of irrigation systems

Table 5.1. Annual irrigation supply to Wonjo-Shoa Scheme

Year	2004	2005	2006	3007	2008	2009	2010	Average
Annual supply, Mm^3	123	117	119	131	112	126	122	121.4

Table 5.2. Irrigation demand of Wonji-Shoa Scheme

Month	Jan	Feb	Mar	Apr	May	Jun	July	Aug	Sep	Oct	Nov	Dec	Total
Demand (l/s/h)	0.38	0.33	0.33	0.39	0.46	0.52	0.21	0.18	0.39	0.48	0.52	0.45	
Irrigated area, ha	6,000	6,000	6,000	6,000	6,000	6,000	6,000	6,000	6,000	6,000	6,000	6,000	
Flow rate (Q), m^3/s	2.28	1.98	1.98	2.34	2.76	3.12	1.26	1.08	2.34	2.88	3.12	2.7	
Volume, Mm^3	6.1	5.3	5.3	6.3	7.4	8.4	3.4	2.9	6.3	7.7	8.4	7.2	80.3

Annual irrigation demand of Wonji-Shoa Scheme is about 80 Mm³. Average measured annual irrigation supply is 121 Mm³. The system is closed for two months (July and august) for annual maintenance. Moreover, these are the two rainy months, and it is assumed that rainfall meets the water requirements. However, there is still substantial irrigation demand during these rainy months. Annually, there is an excess supply of about 51%, accounting for 41 Mm³ of surplus. Figure 5.3 shows monthly irrigation supply versus demand. This amount of water if diverted to new irrigation systems under development, on the downstream of Wonji-Shoa Scheme, it could have been possible to irrigate as much as 3,000 ha of extra land. This is very important, because since recently the competition for water in the basin has been intensified as result of enormous expansion of irrigated land and municipal water needs.

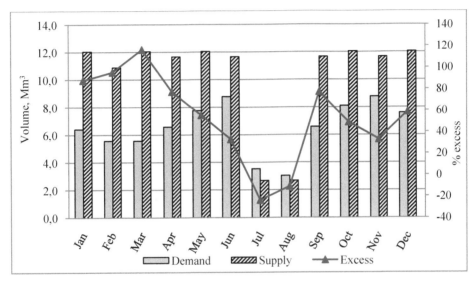

Figure 5.3. Monthly irrigation supply, demand and excess at Wonji-Shoa Scheme

5.3   Water delivery performance

Water delivery performance levels were evaluated on spatial and temporal scales. The spatial indicators are meant for the water delivery performance over a region R; while the temporal indicators are for the water delivery performance in time T.

5.3.1 Spatial performance indicators

Relative delivery, efficiency and equity were considered for spatial performance. The relative delivery (Figure 5.4) shows that the spatially averaged adequacy (relative delivery) of all 16 offtakes for each month was higher than unity for both 2012 and 2013. Average relative delivery values ranged from 1.06 to 1.19, merely indicating 6 to 19% excess water delivery at field levels. The months January, February and March are months of low flow in the river, and there is a consequent low river stage. However, pumping rate at Wonji-Shoa Scheme remained fairly uniform, with acceptable monthly fluctuations. So, it could be adequately assumed that delivery performance over these three months will describe the lowest possible adequacy performance, and hence will address whole year. The monthly variations in relative delivery are not significant, because the management adopts a fairly uniform rate of pumping. Moreover, there are little variations in the monthly irrigation requirements.

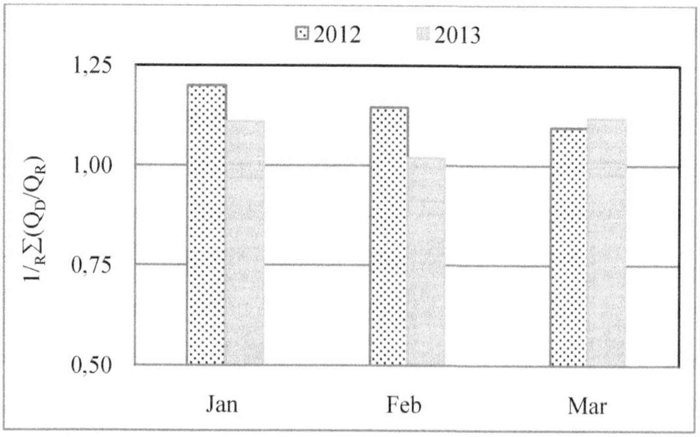

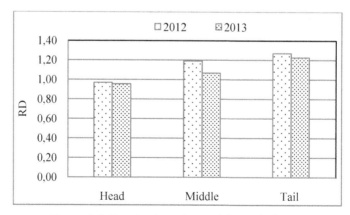

Figure 5.4. Spatial relative delivery indicator

Aggregate relative delivery (Allen *et al.*) at the head, middle and tail reach offtakes is depicted in Figure 5.5 for 2012 and 2013. The value increases from head to tail offtakes. Water delivery to the head reach offtakes is almost as per the field demands, as indicated by a relative delivery value of nearly unity; while tail offtakes deliver the largest supply. This situation could be explained by two factors. First, it is attributed to the sensitivity of flow control structures at offtake points. Offtake structures are overflow and underflow type and water level control structures in the parent canals are all underflow vertical gates. Sensitivity (S) of discharge to changes in water levels (ΔH) is higher for an overflow than underflow structure. The flexibility (F) at offtake structures is lower than one for offtakes in the head reach and higher than one for offtakes in the middle and tail reaches in general. As a result, an increase in supply in the parent canal (right branch) would cause more excesses at the tail reach offtakes. Secondly, water level regulators are customarily operated at fully open conditions, which make excess flow to run to the downstream, eventually over-supplying tail offtakes.

Figure 5.5. Reach wise relative delivery indicator

The monthly spatial efficiency indicator for the two years is shown in Figure 5.6. The values show that the efficiency of water application at tertiary levels is rated as 'good' (higher than 0.85), except a slightly lower value for January 2012. This is an aggregated efficiency for the whole of 16 offtakes. It could be concluded that the efficiency of water use at tertiary levels is adequate as a whole for the secondary canal. However, reach wise efficiency indicators depict inferior efficiency at the tail offtakes (Figure 5.6). Aggregated efficiency levels at the tail reach are rated as 'fair', while that at the head are 'good'.

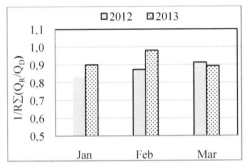
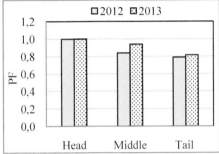

a. Spatial efficiency indicator   b. Reach wise efficiency indicator
Figure 5.6. Efficiency indicators

Equity was evaluated over three months for 2012 and 2013 (Figure 5.7). The equity indicator P_E for the whole offtakes shows that the equity levels rate as 'fair' in the performance standard. However, it is also evident that equity levels are close to 'poor'. Generally, at Wonji-Shoa Scheme, water is pumped in excess of supply. Out of about 51% excess water pumped, only 20% is being diverted to tertiary offtakes. Of this amount, excess deliveries at tail offtakes take the highest share; while that at head offtakes take a lower share. Delivery to head reach offtakes, actually does not contribute to tertiary losses in the scheme. Still, equity at tertiary levels is fairly acceptable, indicating the greatest room for saving irrigation water lies at the main system level (seepage, canal overflows, operational loses, drainage, tail runoff, etc).

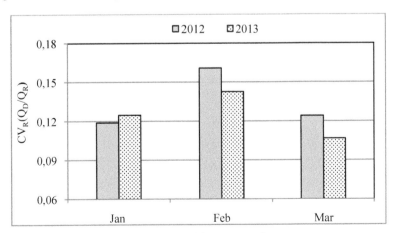

Figure 5.7. Monthly equity indicator

5.3.2 Temporal performance indicators

Temporal relative delivery (Figure 5.8) depicts the temporally aggregated relative delivery of individual offtakes. There is a steady increase in relative delivery values form the head to tail offtakes. The classical assumption that gravity irrigation schemes favour head offtakes in terms of water delivery does not apply for the case Wonji-Shoa Scheme. It is found that institutional arrangements for water management and the typology of the scheme are factors to consider related to this matter. At Wonjo-Shoa Scheme, there are no individual farmers, and there is one central management responsible for the whole scheme. As such, in this scheme 'informal' interventions for unauthorized water diversions, which could be expected from large-scale schemes with individual farmers are absent.

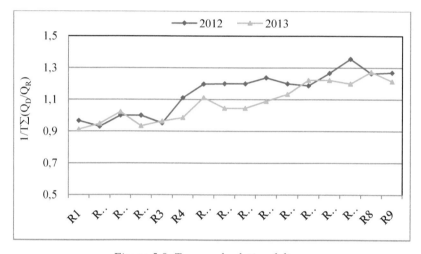

Figure 5.8. Temporal relative delivery

Temporally aggregated efficiency indicator for individual offtakes is shown in Figure 5.9. Offtake temporal efficiency indicators steadily decrease in the downstream direction along the secondary canal.

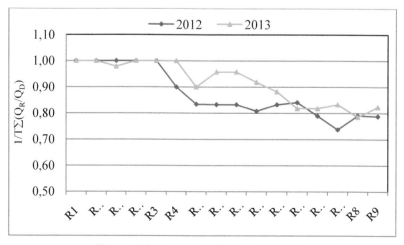

Figure 5.9. Temporal efficiency indicator

Dependability of water delivery at offtakes within each reach is shown in Figure 5.10. This as was discussed, encompasses predictability and variability; where the former refers to reliability of the timing and the later to the reliability of the amount. Flow of Awash River at that reach is highly predictable, because this scheme is located in the Upper Valley, where there are developments on the upstream. As a result of consistent over pumping, reliability of water delivery in each reach remains 'good'.

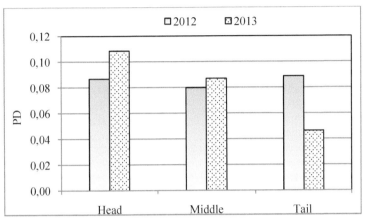

Figure 5.10. Reach wise dependability indicator

5.3.3 Overall water delivery indicators

The overall water delivery indicators for 2012 and 2013 are given in Table 5.3. Regardless of the average excess relative deliveries in each year, the aggregate efficiency, P_F in each year is adequate. This shows that the vast majority of over supplied water is lost before the water reaches in the field (seepage from storage reservoirs, seepage in the conveyance system, operational losses, and tail saline groundwater). Overall, the equity of supply is 'fair' and the dependability rated as 'good'.

Table 5.3. Overall water delivery indicators at Wonji-Shoa Scheme

	P_A	P_F	P_E	P_D
2012	1.14	0.89	0.13	0.08
2013	1.08	0.94	0.12	0.08

5.4   Options for improving water delivery performance

This study showed that operational measures can make a significant improvement in hydraulic performance of Wonji-Shoa Scheme. Based on the results of this study on the water delivery performance of Wonji-Shoa Scheme, the following operational water management options were proposed for improving the hydraulic performance.

5.4.1 Reduced night time pumping

Uniform rate of pumping over 24 hours to the scheme causes over supply in the existing operation. Six pumps with design capacity of 730 l/s each, and two pumps with a capacity of 390 l/s each operate 24 hours a day. In practice, the amount of water required is by far less than what is being pumped, except for July and August, during which the system is closed for annual maintenance. A pumping schedule in which $Q_d = 3.5$ m^3/s during 12 hours of a day, and Q_n reduced to 1.5 m^3/s over the rest 12 hours would adequately supply with enough water. This schedule suffices water requirement during the whole year. On the other hand, a constant rate of pumping at 1 m^3/s over 24 hours is sufficient during July and August.

5.4.2 *Proportional division structures*

The hydraulic characteristics of flow control structures (offtakes and water level control) play a vital role in water distribution and delivery. The flexibility of offtake structures determines water distribution. Offtakes with overflow structures combined with under flow sluice gates as water level regulators generally result in flexibility higher than one. On the other hand, a combination of offtakes with underflow gates and overflow water level regulators result in flexibility less than one. The combination of structures at Wonji-Shoa Scheme oversupplies tail offtakes. Proportional structures ensure proportional division of water under all conditions of supply (excess and shortage). Moreover, aging of structures is a major constraint to effective water management. Installation of proportional division structures at offtakes would improve water delivery equity.

5.4.3 *Flow measurement*

Flow measurement is almost totally absent at Wonji-Shoa Scheme except at 2 major secondary division points. Flow measurement is not being made at offtake structures due to malfunctioning of measurement facilities. At several of these offtakes, measurement facilities have been demolished. Installation of flow measurement structure at each offtake, and periodic re-calibration using current-metering is recommended. Vertical sluice gates and overflow gates could be used both as flow control and measurement facility with attached graduated staff gauge relating stages to discharges. Effect of sedimentation on operation in case of Wonji-Shoa Scheme is not serious as water is supplied from the river by pumping. Moreover, a settling basin and a sediment flushing structure at the head of the main canal substantially reduce further sediment flow into the system. Thus, re-calibration of measuring structures every 2 to 3 years would suffice.

5.5   Conclusion

The existing irrigation water management at Wonji-Shoa Scheme is not adequate with regard to its long-term sustainability and an increasing need for saving irrigation water. Annually, there is a surplus of about 51% being pumped into the irrigation scheme, of which about 40% of the demand (80% of excess) seeps in the main system and storage ponds, drained into escapes and salty tail waters. Percolation losses at field levels account for only 10% of the demand (20% of the total excess diversion). The current water management at Wonji-Shoa Large-scale Irrigation Scheme needs critical attention for two main reasons relate to its sustainability and efficient water use. First, the two environmental issues (waterlogging and salinization) have been greatly challenging the sustainability of the scheme. In over more than 50% of the command area, groundwater levels are within 1 m depth below ground surface. The salinity of shallow groundwater is high (as high as 2 dS/m), which is posing a serious soil salinization in significant portion of the scheme area. Secondly, multi-sectoral water demands in the basin are escalating. These increasing demands including huge large-scale and small-scale large irrigation expansion and rising domestic and industrial water needs in the basin would not allow the irrigation water management tradition and practice at Wonji-Shoa Scheme to continue as it was in the past.

6 HYDRAULIC PERFORMANCE EVALUATION OF METAHARA LARGE-SCALE IRRIGATION SCHEME

6.1 Introduction

The Awash River Basin is a water short region, and is the only source of fresh water in the region for all uses including agriculture, municipal and livestock. Though groundwater has been an alternative source of water, its poor quality prohibits its further development for irrigation and domestic uses. Due to the region's warm climate, about 70% of rainfall in the basin is lost by evapotranspiration (Taddese et al., 2010). On the downstream of Metahara Scheme, there are a number of irrigation schemes supplied from Awash River. These schemes face water shortages particularly during low flows. Moreover, recently there has been vast expansion of irrigated agriculture in Awash River Basin with an objective of implementing sedentary farming for the pastoralist community. This in turn has intensified the competition for water of Awash River and necessitated a more integrated water management of the basin.

Metahara Scheme is blamed for excessive water diversion and wastage to saline swamps at its tail ends. With an objective of water saving and ensuring its sustainability, this chapter evaluates irrigation supply and demand of the scheme, and assesses the water delivery performance of it. There were no sound studies in the past on this scheme to address scheme water demand versus diversion. Moreover, water delivery performance monitoring within the scheme is absent. This study is therefore a useful start for improving overall irrigation water management, distribution and delivery in the scheme.

6.2 Water acquisition, conveyance and delivery

The Metahara Scheme has two parts called 'Main System' (8,000 ha) and 'Abadir' (3,500 ha). There are separate diversion intake structures about 3 km apart for the two sub systems. Abadir sub- system, located in left bank is supplied by the upstream intake, while the Main System on the right bank is supplied by the Main intake structure. This study focuses on the water delivery performance of the Main System.

Diverted water via a diversion weir for the Main System is being conveyed and distributed in an extensive network of open canals. Exclusively manual structures (vertical sluice gates and overflow Romijn weirs) are used for flow control. The canal system consists of main canal, secondary canals, lateral canals and sub-laterals (tertiary canals). The layout of the irrigation canal and offtakes of the Main System (main canal and part of a secondary canal) is shown in Figure 6.1. The water distribution system is continuous at the main level and rotational in the tertiary level. The flow of a sub-lateral (tertiary canal) is 200 l/s, which is being handled in rotation at different outlets, by an irrigation crew of three persons.

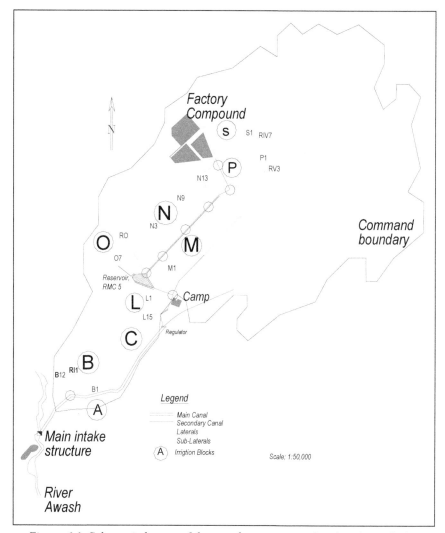

Figure 6.1. Schematic layout of the canal system considered and its offtakes

6.3   Data collection for water delivery/hydraulic performance

Climate data, data on location and condition of flow control and offtake structures and discharge are the main data required. Climate data were required for determination of evapotranspiration, and hence determination of scheme water and irrigation demand. Metahara Scheme has its own weather station, where daily records of data on rainfall, temperature, wind speed, relative humidity and sunshine hours are available for over 40 years. For the canal with a total length of 11 km (about 6 km main canal and 5 km secondary canal), location of each flow control and offtake structure, along with their salient dimensions and hydraulic features were collected. Flow measurement is the core of water delivery performance assessment, as this depends in this case on routinely monitored irrigation flow data. Flows at 15 lateral and sub-lateral offtakes along the canal considered, were monitored for 2012 and 2013 over the three critical months (January, February and March). These are the months with low Awash River flows. Irrigation demands during these months are also lower; however the low flow of the river is these months is more important (constraining) to the water delivery performance than the higher water demands during other months. Delivered flows (Q_D) were measured on daily basis, using current meters and in some cases using V-notch weirs, from which the average monthly flows were determined. The required (design) flow (Q_R) in each offtake depends on the area it irrigates; it is 200 l/s for sub-laterals, and depends on the number of tertiary units fed for lateral canals.

6.4   Water delivery performance indicators

The water delivery (hydraulic) performance indicators used in Chapter 5 were applied for this scheme as well. Adequacy (relative delivery) indicator (P_A), efficiency indicator (P_F), equity indicator (P_E) and dependability indicator (P_D) Molden and Gates (1990) were the main indicators of internal water delivery performance used in this study. Indicators were determined for each offtake, as well as for a group of head, middle and tail offakes in order to describe any spatial and temporal variations.

6.5   Irrigation water demand/supply

6.5.1   Monthly and annual irrigation demand

Monthly irrigation demand was determined as the net difference between monthly water demand and effective rainfall plus the leaching requirement (*LR*) (Table 6.1). CROPWAT (Swennenhuis, 2006) was used for water demand determination from climatic, crop and cropping pattern data. The average salinity of irrigation water (*ECw*) is 420 µS/cm, while the root zone crop salt tolerance level for sugarcane is about 1,700 µS/cm (Tanji and Kielen, 2002). For this condition, a leaching fraction of 0.05 was assumed adequate.

6.5.2   Monthly and annual total irrigation supply

Records of daily and hence monthly irrigation supplies are being taken by the Estate with a stage-discharge relation established at the head reach of the main canal. Annual irrigation supplies for 5 years (2006 to 2010) were determined as a sum of daily and hence, monthly supplies from the *Q-h* relation. The 5 years average annual water diversion is 190 Mm³, which exceeds demand by 24%, and this means 37 Mm³ of excess water diversion annually. While there is a claim of shortage in significant parts of the scheme, excess diversion apparently means wastage in the conveyance, storage and distribution systems. The main rainy season in the area extends from July to September; however the rainfall is very low to meet the water requirement in these months. While there is excess diversion at the intake during the other months of the year, August and September are months with field demands exceeding diversion. The river flow during these months is sufficiently high. However the managers assume that the rainfall would meet the demand while it would not. Moreover, during these two months major annual maintenance takes place. Monthly average irrigation water supply (diversion) in comparison with demand, along with percent excess/shortage is shown in Figure 6.2.

Table 6.1. Monthly and annual irrigation demand of Metahara Scheme

Item	Jan	Feb	Mar	Apr	May	Jun	Jul	Aug	Sep	Oct	Nov	Dec	Total
NIR, mm/month	136	110	123	134	193	241	157	132	174	182	175	152	1,910
NIR, l/s/h	0.51	0.45	0.46	0.52	0.72	0.93	0.59	0.49	0.67	0.68	0.68	0.57	
Q, m^3/s	4.1	3.6	3.7	4.1	5.8	7.4	4.7	3.9	5.4	5.4	5.4	4.5	
Irrigation demand, Mm^3	10.5	8.8	9.8	10.7	15.4	19.3	12.6	10.6	13.9	14.6	14.0	12.2	152

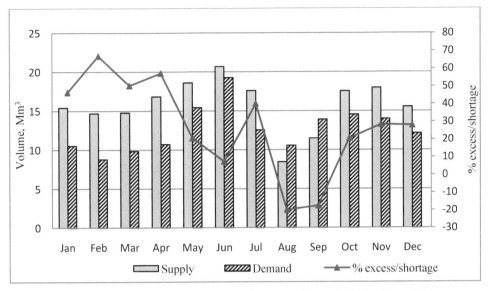

Figure 6.2. Monthly irrigation water supply versus demand of Metahara Scheme

6.6 Water delivery equity performance at scheme level

At Metahara Scheme, misunderstanding the hydraulic characteristics and hence inadequate operation of flow control and measurement structures are found to play a significant role for non-optimal water distribution within the scheme. There is practically no monitoring mechanism that verifies flow measurement at offtakes. The flow control structures at tertiary offtakes are either Romijn weirs (overflow weirs) or sluice gates; while those at secondary offtakes are sluice gates. It was found that there are significant variations between the actual flows through the structures and the flows assumed by operators. So, continuous monitoring and calibration of control structures was found to be critical for improving the operation and enhancing the water delivery performance.

The non-uniformity of water distribution in the scheme was visible both spatially and temporally. To have a first insight on the degree of non-uniformity of water distribution within the scheme in general, spatial equity levels of water deliveries at offtakes were determined for the scheme as a whole. For this, 10 tertiary (sub-lateral) offtakes were randomly considered from the head, middle and tail reaches all over the scheme. Actual daily flows were monitored over a period of six months (January-June, 2011) at these offtakes (Table 6.2), which were then compared to the intended flow of 200 l/s for each. Spatial coefficient of variation (CV) was used as an indicator of spatial equity of delivery in the scheme as a whole. The CV of the delivery performance ratio (DPR) is a useful indicator for equity at different levels. Tariq et al. (2004) and Vos (2005) applied CV to evaluate the level of variability of irrigation supplies at offtakes and water delivery performance at main, secondary and tertiary levels respectively. For tertiary offtakes, a target CV of 10% is assumed practically acceptable in systems with good water management. Nevertheless, CV of 32% was obtained in this case, which is much more than the target. Apparently, this gives an insight that the water distribution among tertiary offtakes over the whole scheme is remarkably non-equitable. Table 6.2 shows that the measured average flows at eight offtakes are in short of the intended flows. However, there has been significant amount of excess irrigation water diversion at the source unlike shortages at offtakes. Apparently, the diverted excess water accounts for percolation and seepage losses and operational and drainage losses within and at the tail ends of the distribution systems.

Table 6.2. Measured average flows (6 months) and deviations at selected tertiary offtakes

Offtake	1	2	3	4	5	6	7	8	9	10
Measured average Q, l/s	159	189	109	200	124	69	141	222	101	169
Deviation, l/s	-41	-11	-91	0	-76	-131	-59	22	-99	-31
Deviation, %	-21	-6	-46	0	-38	-66	-30	11	-50	-16

6.7   Hydrodynamic characterization of the canal system

Of a total of 15 offtakes considered on the canal, the first 6 are in the head reach, next 5 in the middle reach and last 4 in the tail reach. There is a night storage reservoir in the system at a distance of 6 km from the headwork. Offtakes from offtake O_7 and downstream are supplied from this reservoir. There is a gated underflow water level regulator at offtakes, except for B12, B1, RI1 and N8. The design of the system is based on 9 hours of irrigation per day (6 AM - 3 PM), with constant offtake flows. However, with extensive field observations and measurements, it was found that the design criteria are not followed in the day to day management of the irrigation system. Sound operation rules for the offtakes are lacking except for some major offtakes. Often operators use their own presumptions of operation, which causes large fluctuations in water levels and discharges during a day. This has in turn caused inefficient water use and non-uniform water distribution in the scheme. Table 6.3 shows some of the features of the offtakes in the system considered.

The flow in the system during night hours is minimal. When opened early in the morning, the system responds slowly to assume design conditions. As such, the flows at offtakes increase gradually and on average tops design discharges 3 to 4 hours after opening. There occurs significant excess flow for 2 to 3 hours on average and offtake flows start to decline once again with depletion of water levels in supply canals and storage reservoirs. Sensitivity of offtake structures in many cases is greater than unity, which contributes to lots of tail end drainage. During excess flows, irrigators could not have effective control on the water, causing significant field and off-farm losses. While flows as high as 200% of the offtake design discharges were occasionally observed during the day, as low as only 40% of the design offtake discharges were also observed in the monitoring process. Lag-time and the unsteady state situations of the flow control system are the major causes of these fluctuations and losses. Particularly, the first two offtakes in the middle reach are located just at the outlet of a reservoir; and hence have an insignificant lag-time. However, the flow starts to decline significantly 2 to 3 hours after opening. The hydrodynamic behaviour has distinct effects on the on the offtake flows at the head, middle and tail reaches of the system in general (Figure 6.3). It shows that middle reach offtakes, fed directly from a reservoir have the largest fluctuation as a result of mis-operation and water level fluctuation (perturbation) in the parent canal.

Table 6.3. Features of offtakes in the three reaches of the canal system

Location	Head reach						Middle reach					Tail reach			
Offtake name	B12	B1	RI1	L15	L1	M1	O7	RO	N4	N8	N13	P1	RV3	RIV7	S1
Q_R (m³/s)	0.2	0.2	0.8	0.6	0.2	0.2	0.2	1.0	0.2	0.2	0.2	0.2	0.6	0.8	0.2
Offtake type	S	S	S	L	S	S	S	L	S	S	S	S	L	L	S
Flow type	U	U	U	U	U	O	O	U	O	O	O	O	O	U	O
WL regulator	No	No	No	Yes	Yes	Yes	Yes	Yes	Yes	No	Yes	Yes	Yes	Yes	Yes
Operational lag-time (hours)	1h	1h	1h	4h	4h	4h	0h	0h	1h	2h	3h	4h	4h	4h	4h
Perturbation level	M	M	M	M	M	M	C	C	A	A	A	M	M	M	M

S = Sub-lateral L = Lateral U = Under flow O = Over flow M = Minor A = Average C = Considerable

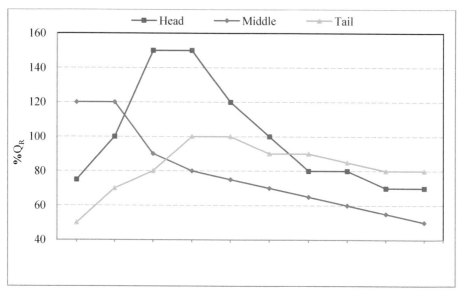

Figure 6.3 Average measured fluctuations of discharges in a day at head, middle and tail offtakes for Metahara Scheme

6.8   Evaluation of the water delivery performance

The water delivery performance of the system was evaluated based on flows monitored at each offtake. The intended flow of sub-lateral canals is 200 l/s, while that of laterals is variable based on the number of tertiary offtakes it supplies. For each of the 15 offtakes, the monthly average actual delivered flows for a period of 3 months were determined from daily flow measurements by current-metering for 2012 and 2013. Flow measurements were made twice a day to capture daily temporal fluctuations.

Recently, with the expansion of new irrigation schemes in the Awash River Basin, there has been an increasing need for saving water in existing schemes. To this end, the Awash River Basin Authority is working to achieve a more effective and integrated water resources management that would enhance equity in sharing water in the basin. Regardless of its excess water diversion, Metahara Scheme claims of water shortages itself. However, the analysis on water diversion-demand (based on measured water diversion and calculated irrigation demand) confirmed that excess diversion amounts to about 24%. But where is the excess water diverted lost; as seepage in the conveyance and distribution, in the field or as tail drainage? How does irrigation water delivery behave at various offtakes? For this, both spatial and temporal delivery performance needs to be evaluated at the offtakes. The results of performance evaluation are in the following sections.

6.8.1   Spatial performance indicators

Spatial performance indicators are spatially-averaged values of indicators of water delivery performance of all offtakes for different periods of time. Spatial values of indicators were evaluated for adequacy, efficiency and equity for three months of 2012 and 2013 (Figure 6.4). Spatial adequacy levels for 2012 were relatively inferior ranging from 'fair' to good' over the months. Adequacy levels were superior for each month in 2013 and there is insignificant monthly variation. Despite lower values for February and

March of 2012, spatial adequacy levels are acceptable. On the other hand monthly aggregated efficiency indicators are shown in Figure 6.4. These values concern the tendency of the whole system for saving water on the downstream of the offtakes. Values of nearly unity depict that offtakes deliver nearly the required amount water. This also means that there is no significant water loss at tertiary and field levels.

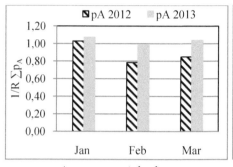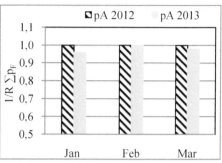

a. *Average spatial adequacy, p_A*   b. *Average spatial efficiency, p_A*

Figure 6.4. Spatial adequacy and efficiency indicators

The spatial coefficient of variation of water delivery (Figure 6.5) shows the degree of spatial equity (uniformity) of water delivery to all offtakes over 3 separate months. The equity level was 'poor' for February and March in 2012, while it was 'fair' for each month in 2013. It is evident that there is a significant fluctuation in equity levels from one year to the other for the same month. Moreover, unlike considerable monthly fluctuations in equity levels, the offtakes design flows remain practically unchanged over the months. Absence of sound and demand responsive operation rules is responsible for the inequity.

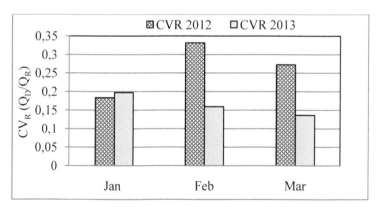

Figure 6.5. Average spatial coefficient of variation, CV_R

6.8.2 Temporal performance indicators

Temporal values of performance indicators were evaluated for adequacy, efficiency and dependability. The indicators were assessed for each offtake and reach-wise (head, middle, tail) for 2012 and 2013. The temporal indicators show that the level of adequacy of water delivery is generally more than adequate at head and tail reach offtakes, both for 2012 and 2013 (Figure 6.6). However, adequacy got inferior at the middle offtakes, which could be explained by the following two factors. First, there is inadequate of operation of the night storage reservoir, which causes significant temporal fluctuations in water stage in the reservoir outlet canal, from which middle offtakes are supplied (Figure 6.7). Second, all the sub-lateral offtakes in the middle reach except one lateral, have overflow structures, while the water level regulators are underflow. Hence, the offtakes are hyper-proportional. This means that any change in the flow or water level in the parent canal generates relatively larger perturbations in the flow of these offtakes.

The temporal average efficiency indicator for each offtake is shown in Figure 6.6, in which the level of performance ranges from 'fair' to 'good'. Reach-wise, efficiency levels are nearly unity at each offtake in the middle reach, showing that there is no excess water delivery at these offtakes. Efficiency values in general indicate that water losses in the system at sub-lateral (tertiary) levels are very little, and they give insight that excess diversion is lost in the main system, into drainage within the system, and at system tail end.

The temporal coefficients of variation of water delivery, which indicate the dependability of supply over 3 months for 2012 and 2013 are shown in Figure 6.8. It is evident that for both years, the supply was more predictable for tail reach offtakes than for head and middle. Dependability of delivery at head and middle offtakes significantly varied from one year to the other as well as from one offtake to the other. Particularly, dependability was very poor in 2012 at 5 offtakes in the head and middle reaches. Flow perturbation due to inadequate operation is more absorbed by head and middle offtakes due to the hydrodynamic characteristics and nature of offtakes and cross regulators.

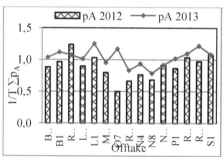
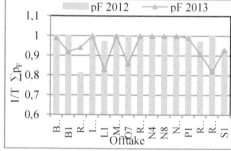

a. *Average temporal point adequacy, p_A*   b. *Average temporal point efficiency, p_F*

Figure 6.6. Temporal adequacy and efficiency indicators

Figure 6.7. Low stage in a parent canal behind offtakes in the middle reach

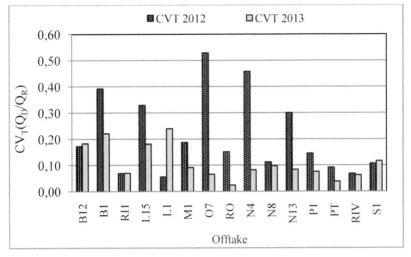

Figure 6.8. Average temporal coefficient of variation, CV_T

6.8.3 Reach averaged performance indicators

In addition to spatial and temporal indicators of water delivery, reach-aggregated indicators enable to understand the water delivery performance better. Average indicators of adequacy and efficiency in each reach are given in Figure 6.9. It indicates that the reach aggregated water delivery exceeds demand by 7 and 10% at the head and tail reaches for 2013. Adequacy (Sakthivadivel *et al.*) for 2013 within each reach rated 'good' (sufficient or excess); while for 2012, P_A got 'poor' for the middle reach. In each year, P_A value was inferior in the middle reach. This is also clear from Figure 6.6 of individual offtake adequacy values, while in this case aggregated values are presented. Aggregated reach P_F values were greater than 0.90 for each reach and year, indicating good performance. Though there is iniquity across the reaches, efficiency at tertiary levels is all acceptable.

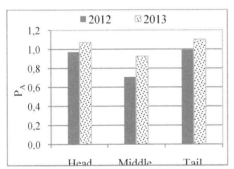
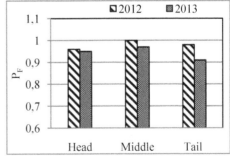

a. Relative delivery indicator (adequacy), P_A b. Efficiency indicator, P_F
Figure 6.9. Reach-wise relative delivery and efficiency indicators

6.8.4 Overall performance indicators

Overall performance values indicate the average water delivery performance of 15 offtakes over a period of three months for 2012 and 2013 (Table 6.4). Overall Adequacy

(Sakthivadivel *et al.*) values indicate that the water delivery was 'fair' and 'good' for 2012 and 2013 respectively. Efficiency (P_F) implies that the overall offtake efficiency was 'good' in each year as per the delivery standard of Molden and Gates (1990), Renault and Wahaj (2007). Unlike water shortage claims at this scheme, the water delivery to offtakes in terms of quantity at sub-lateral and lateral levels is fairly adequate on average. Overall offtake efficiency is indeed superior. Average overall equity (P_E) of delivery was 'fair' in each year. Average dependability (P_D) was 'fair' and 'good' for 2012 and 2013 respectively. Generally, equity needs more attention than other indicators regarding water delivery at lateral and sub-lateral levels as a whole.

Table 6.4. Average overall values of water delivery performance indicators

Year	P_A	Level	P_F	Level	P_E	Level	P_D	Level
2012	0.89	Fair	0.98	Good	21	Fair	20	Fair
2013	1.03	Good	0.94	Good	14	Fair	10	Good/Fair

6.9   Operational measures for improving water delivery performance

The major cause of inferior water delivery performance at Metahara Scheme is lack of good understanding of the complex hydrodynamic characterises and the associated lack of adequate operation that could cope with it. Although, a hydrodynamic modelling in Chapter 5 will provide more detailed aspects on the operation of different components of the scheme, the following operational measures have been proposed based on this empirical study.

6.9.1   Operation of the reservoir RMC5

The largest night storage reservoir (RMC 5) is located at a chainage of 7+00 in the canal system considered. It supplies the middle and tail offtakes, along this secondary canal. The operation of this reservoir plays a vital role in the water distribution and delivery on its downstream. Middle reach offtakes are just on the downstream of it, and are being supplied from a canal offtaking directly from the reservoir. The outlet of the reservoir is controlled by a sluice gate with a maximum opening height of 1.8 m. In the existing operation, the outlet sluice gate is in general fully opened during irrigation hours. Because of serious sedimentation of the reservoir, its supply sustains only for the first 3 hours after the sluices are fully opened at 6 AM. Afterwards, the water level in the offtaking canal significantly drops, and causes large flow reductions in the offtakes in the immediate downstream. Frequent operation of the control sluice at least 3 times during a day will help to address the problem of inequity and middle reach inadequacy. Progressive reduction in the opening height of the gates during the daily operations is recommended.

6.9.2   Operation considering sedimentation

Apart from in reservoirs, there is also a severe sedimentation of conveyance and distribution canals. Sediment exclusion and flushing facilities are absent, and a huge sedimentation ends up in the system. Sedimentation is indeed one of the major causes of ineffective operation at Metahara Scheme. In the head reach, water is delivered directly from the main canal, whose bed level is lower than the crest levels of offtake structures. As such sedimentation has little or no impact on offtake flows, because bed sedimentation of the main canal still remains below the offtake levels. Thus, fully open operation of head reach offtakes enables them draw adequate and often excess, in an attempt to divert sufficient supplies. The combined effect of this and sedimentation of reservoirs is eventually reflected significantly reducing the flows in the middle reach offtakes. Operational options are:

- *partial opening of head reach offtakes*. Head reach offatkes have to be partially opened in order to ensure a better equity and adequacy along the canal. As these offtakes draw water from a deeper canal, water level fluctuation in the canal causes little perturbation on offtake discharges;
- *partial opening of reservoir RMC5 outlet*. In order to keep water levels fluctuation in the offtaking canal from the reservoir to minimum, frequent reservoir outlet operation with partial opening is useful. Moreover, this reduces runoff discharge at the downstream ends. Successively reducing gate opening heights of 1.8, 1.5, and 1.2 m for three hours each is recommended during 9 hours irrigation per day;
- *reduced opening of cross regulators in the middle reach offtakes*. Reducing the opening of water level regulators at offtakes fed from the storage reservoirs would help to maintain a more or less higher water level in the parent canal. This enables middle offtakes to draw a sustained and uniform flow rate during the irrigation hours. It also hinders excess flow down the system, which would have been eventually discharged to saline waterlogged areas at the tail ends.

6.9.3 Varied operation time for offtakes

In the existing operation, advance time of water, while it is quite basic for a gravity irrigation system, is given no attention. Water takes on average about 2 hours to advance from the major storage reservoir (RMC5) to tail offtakes. However, no operational measures exist for taking the advance time into consideration. All the offtakes along the canal are being operated uniformly at similar times during a day (6 AM and 3 PM). Sudden closure of tail offtakes at 3 PM not only causes huge tail reach inundation, but also reduces the delivery time of tail offtakes. Actually tail enders had better adequacy than middle, but it is because of higher flow rates and not due to longer duration of supply. Postponing the time of closure of tail offtakes from 3 PM to 5 PM helps to ensure a more equitable delivery time throughout and to minimize tail flooding.

6.9.4 Flow measurement and regular recalibration

Recalibration of flow measurement structures at branching points and offtakes is absent. Sedimentation greatly affects the calibrated discharges of structures by reducing the flow section of offtakes. The measurement structures were installed and calibrated 2 to 3 decades ago. It would be unrealistic to assume offtake flows measured with these structures in view of the huge sedimentation. However, operators still make use of these structures for measurement without any re-calibration. Often, offtakes operate under submerged flow conditions instead of free flow due to sedimentation of offtaking canals (Figure 6.10). Even the existing old calibration takes no account of the effect of submergence at all. It was confirmed by field flow monitoring that on average, these offtake flow measuring structures overestimate the actual flows by nearly 25%. Two distinct calibrations, one for free flow and another for submerged flow are actually required for reasonable flow measurement. As it will be unrealistic to consider flow calibration for submerged flow conditions at Metahara, it is recommended to regularly calibrate the measuring structures every 3-4 years for a realistic flow measurement. It is well understood that particularly the flows in sub-lateral (tertiary canals) turns into submerged flow in less than 1 year time after canal cleaning. Hence, an adjustment is recommended to the measured offtake discharges whenever calibration is older than 1 year and whenever the offtake canals are not properly maintained or cleaned. The flow adjustments required depends on the degree of sedimentation and specific condition of the offtakes.

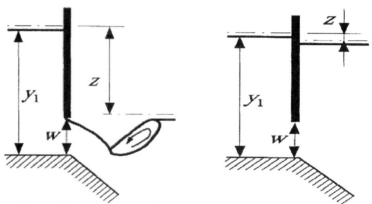

a. Free orifice flow   b. Submerged (conveyance) under flow (Ankum, 2002)
Figure 6.10. Free and submerged underflow conditions for a vertical gate applicable to the case of Metahara Scheme

6.10   Conclusion

In the existing condition, the operation rule for diversion of irrigation water at the main intake of Metahara Scheme is not adequate. The rule is not sound in that demand and supply are poorly matched. The diversion was in short of demand in August and September, during which the supply is being cut-off for canal maintenance. As August and September are rainy months, managers assume rainfall would meet the water demand. However, while Metahara Scheme is in a semi-arid region, irrigation is required the whole year. On the other hand, water diversion during the remaining 10 months is excess, accounting for about 24% annual surplus. Water scarcity and hence competition for water in the basin is currently steadily increasing due to expansion of community-based and public irrigation schemes and rising municipal and industrial demands. If could be saved, the excess diversion could have irrigated as much as 1,400 ha of land elsewhere in the basin.

The average water delivery performance to offtakes at tertiary levels is in general more or less adequate in terms of adequacy and efficiency, and hence there is little water loss at field levels. However, the operation of offtakes was not adequate in terms of overall equity. Inequity is directly related to limited knowledge of canal managers and operators on the complex system hydrodynamic characteristics and hydraulic nature of flow control structures. On the other hand, reach wise evaluation showed that inferior offtake water delivery adequacy at middle offtakes is due to inappropriate operation of night storage reservoir and hyper-proportional nature of offtakes. More than 90% of the unintended diversion into the scheme is lost as operational losses at divisions, into drainage within the scheme, as seepage in the distribution, and to saline and waterlogged tail end swamps. Groundwater is saline in the region; and excess percolating water is not recoverable. Moreover, due to rising groundwater table, salinization has already become a critical problem at this scheme, particularly in the tail ends. A demand-responsive operation plan based on meticulous hydraulic analysis of the flow conditions and flow characteristics of control structures instead of operators' presumptions will be required. To this end, a further study on the hydrodynamic analysis and operation plans will be presented in Chapter 7.

7 MODELLING FOR HYDRAULIC PERFORMANCE AND EFFECTIVE OPERATION OF METAHARA LARGE-SCALE IRRIGATION SCHEME

7.1 Objectives of the hydrodynamic modelling

Hydrodynamic models have proven to enable better understanding of the overall hydraulic characteristics of large-scale irrigation schemes and are useful for evaluation of the impacts of different operational options on the hydraulic performance. In Chapter 6, the water delivery performance of Metahara Scheme was evaluated using routinely monitored flows. While this assessment gives a useful clue regarding the existing water distribution and delivery, it is not possible to appraise the effects of various operational measures. Thus, the hydraulic model DUFLOW was calibrated and was used as a tool for better understanding the hydrodynamics and to aid decision making in operation. To this end, first the hydraulic performance of water diversion and delivery under the existing operation was evaluated. Secondly, three alternative operation rules were developed and hydraulic aspects were simulated for each. The performance of each operation was assessed based on the simulation results. The proposed operation measures would improve the hydraulic performance and help gain significant saving on the water diverted from the source. By matching supplies to demands, and reducing percolation, the measures would also contribute to the sustainability by slowing saline water table rise. The results and lessons to be learnt from Metahara Scheme could be extended to improve the operation of other similar schemes.

7.2 System selection, data requirement and model setup

7.2.1 The irrigation network selected

The selected irrigation canal of Metahara Scheme for simulation has a total length of 11 km. It comprises of a main canal of about 6 km and a secondary canal of about 5 km length. At the end of the main canal is a reservoir (RMC5) from which the secondary canal offtakes . A total of 16 offtakes (Figure 7.1), offtaking from the canal were considered for hydraulic performance evaluation.

7.2.2 Data requirement for model setup

Adequate data on water conveyance and distribution infrastructure, flow control structures and their operation, water levels, and discharges, which are prerequisites for adequate hydraulic simulation, were put together for modelling. Field data collected fall in three main categories: 1. data on canal profiles and cross sections; 2. data on flow control structure details and their operation; 3. data on water levels (depths of flow) and offtake discharges. The first two are required for model setup while the third is required for model calibration and validation. Canal profiles were determined for the whole 11 km canal length (main canal and part of secondary canal) by Total Station Surveying equipment (Figure 7.2) and canal cross-section data were measured at fixed intervals by a campaign of canal surveying.

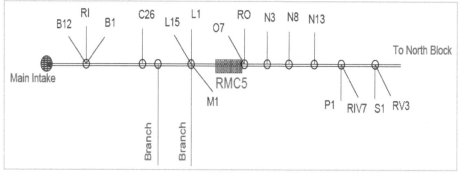

Figure 7.1. Layout of the canal system selected along with its offtakes at Metahara Scheme

Structures such as cross regulators (mostly sluice gates), offtakes and culverts were carefully located, and their dimensions were measured. Moreover, their detailed features such as gate levels, opening heights, maximum opening and their conditions were carefully noted for each structure. The operational features of these structures were also studied. Levels (flow depths) were measured using divers (pressure sensors) installed at selected points along the canal for calibration and validation. Gates of flow control structures were set to pre-set adjustments in order to ensure same conditions for simulations. The flow in offtakes is unsteady due to unsteady nature of flow in the main system. Hence, discharges were measured at same time under same conditions, but on different days for calibration/validation, and thus enabling calibration/validation in space instead of in time. Flow depths and discharges for calibration were measured in February 2012, while for validation in March 2013.

a. *Measuring details of structures*   b. *Canal profile surveying*
Figure 7.2 Surveying details of structures and canal profile

7.3   Hydraulic performance measures

7.3.1 Total water diversion versus demand

The main purpose of a head regulator is to regulate the inflow based on field irrigation demands. In manually-operated schemes, operators often mis-operate regulating structures, such as weirs and gates. An important aspect of hydraulic performance for operation of a headwork is matching demand and supply. This is particularly important in schemes where water saving and efficiency are important. A crucial element for discharge at the diversion is the size of opening of a structure, particularly when the water level could be kept more or less constant. Plots of discharges at various sizes of gate setting openings (flow whirlpool) for the specific structure gives a practical indication of whether there is oversupply or undersupply. Hydraulic simulation makes it easier for discharges and delivered volumes to be simulated and be compared to assumed discharges by the operators for different height of whirlpool.

7.3.2   Performance ratio (relative delivery)

Performance ratio implies the extent to which the intended amount of water is satisfied by the amount of water delivered at an offtake. However, it does not give a clue on the efficiency of water use within the irrigation unit. It is expressed as:

$$PR = \frac{V_{supplied}}{V_{intended}}, PR_{reach} = \Sigma_R \frac{V_s}{V_i}$$ (7.1)

Where V_s is supplied volume (m^3) and V_i is intended volume of delivery (m^3). The intended volume is based on field water demand and the supplied volume can be either measured or simulated.

7.3.3   Operational efficiency

It is well comprehended that efficiency in irrigation is a common term used to describe the water conservation behaviour of different components of an irrigation system. Efficiency can be applied to the whole irrigation scheme, to the conveyance system, to the distribution and off-farm system, to an offtake, or at the field level. However, conveyance efficiency is not suitable for hydrodynamic performance assessment, and it implies seepage and spillage losses (Ankum, 2002). It is given as:

$$OE = \frac{V_{effective}}{V_{supplied}}, OE_{reach} = \Sigma_R \frac{V_e}{V_s}$$ (7.2)

Where V_e is the effective volume (m^3) and V_s is the supplied volume of delivery (m^3). The effective volume of water delivery, the part of supplied water that was effectively controlled and applied to the field, depends on the flow rate and timing of application. For instance, in an intermittent irrigation water delivery to an offtake, it takes some time for the flow to assume design states during which loses occur. Moreover, when cutting the offtake flow, losses occur. These losses are called operational losses, and are significant in intermittent offtake flow irrigation systems.

7.3.4   Equity (spatial uniformity)

Equity, an indicator for the delivery of a fair share of water to different parts of a scheme, is a crucial indicator for water saving, irrigation service provision, sustainability, etc. Equity was evaluated by various researchers in the past in relation to the supplied and intended volumes of delivery. However, as the water volume delivered does not necessarily denote its beneficial use, instead of supplied volume, an equity based on an effective volume of delivery given as below is proposed.

$$P_E = CV_R \left(\frac{V_e}{V_i}\right) \tag{7.3}$$

Where V_e and V_i are as defined above.

7.4   Model calibration and validation

Observed flow depths at specific locations along the main (secondary) canal and measured offtake discharges were used for calibration and validation. Flow depths measured every 1 hour at salient locations along the canal whose flow is to be simulated were considered. Flow depth measurements were made for two different gate adjustments of control structures (main head regulator and cross regulators), at different times for calibration and validation.

The flow in the main system is all unsteady due to refilling required for night diversion reduction on daily basis and canal storage depletion. Similarly, offtake discharges are also unsteady and more or less steady discharges attained 3 hours after offtake opening (9 AM) were monitored for calibration/validation. As such, calibration and validation was made in space for offtakes along the canal instead of the unsteady flow for each offtake. For this, the whole canal reach of about 11 km long was divided into two reaches, where offtake flows in each reach were considered independently for calibration/validation. Offtake discharges were measured at same time on different days while keeping similar gate adjustments to make sure that the hydrodynamics remain same. For validation, offtake discharges were measured under a different hydrodynamic condition than calibration. Canal Chezy roughness (C) and discharge coefficients of major structures (C_d) were used as calibration parameters. These were varied until acceptable simulated values of water levels and offtake discharges were obtained. Statistical indicators: root mean square error (RMSE) (normalized), coefficient of residual mass (CRM), and model efficiency (ME) were used as measures of model performance. Mathematically expressed as:

$$RMSE = \sqrt{\frac{\sum_{i=1}^{n}(M_i - S_i)^2}{n}} * \frac{1}{M} \tag{7.4}$$

$$CRM = \frac{\sum_{i=1}^{n} M_i - \sum_{i=1}^{n} S_i}{\sum_{i=1}^{n} M_i} \tag{7.5}$$

$$ME = 1 - \frac{\sum_{i=1}^{n}(M_i - S_i)^2}{\sum_{i=1}^{n}(M_i - M)^2} \tag{7.6}$$

Where M_i = measured values, S_i = simulated values, M = mean of measured values, n = number of measurements. Lower (closer to zero) values of the RMSE denote a more accurate simulation. Dimensionless normalized root mean square error (NRMSE) to the mean of the measured values enables absolute comparison of model performance. The CRM is an indicator of existence of consistent errors in the distribution of all simulated and measured value with no consideration of the measurement order, and its value of zero indicates no bias. The ME is a measure for the accuracy of simulations. Its maximum value of 1.0 indicates perfect match of simulated and measured values.

Offtake discharges and flow depths were measured for calibration and validation. The DUFLOW network schematization window for the plan of the canal with its flow control structures and offtakes is shown in Figure 7.3. Offtake discharges were measured at 14 lateral and sub-lateral offtakes along the canal for this purpose. Two sets of data of offtake discharges were monitored for one for calibration and the other for validation. Offtake flows measured under different hydrodynamic situations were attained by different gate settings of offtakes and control structures. It is not feasible to calibrate based on a time series of measured flows at each offtake independently; because the flow at each offtake is a result of collective hydrodynamic behaviour of the system. Thus, the model was calibrated using average flow rate of each offtake, which were determined from two measurements made in a day (8 AM and 2 PM). In addition to offtake flows, flow depths were also monitored at two salient locations in the canal system for calibration (at chainages of 1+300 and 7+100) for calibration. The first point was in the main canal; while the second point was located in the secondary canal downstream of Reservoir Main Canal (RMC5), just downstream of the reservoir outlet. The flow depths were measured at these points at 1 hour interval, starting at 6 AM when field irrigation started (offtakes were opened) till 3 PM when offtakes were closed.

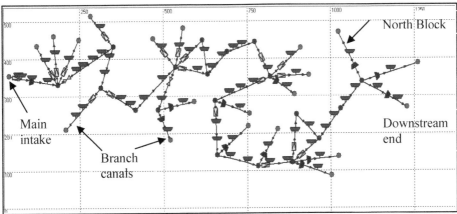

Figure 7.3. Schematization of the canal system and offtakes in the DUFLOW screen

Measured and simulated offtake discharges for calibration and validation are shown in Figures 7.4 and 7.5. Similarly, measured and simulated flow depths at the two locations for calibration are shown in Figures 7.6 and 7.7. The calibrated Chezy roughness coefficient C for lined and unlined canals, and discharge coefficient C_d for underflow and overflow structures are shown in Table 7.1. Statistical indicators for the model performance are shown in Table 7.2. The RMSE for offtake discharge indicates fair model performance for both calibration and validation for the calibrated parameters. However, its performance for water level is good, indicated by low values of NRMSE. However, CRM and ME perform good both for offtake discharge and water levels (calibration and validation).

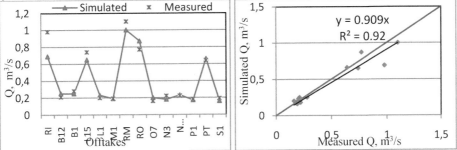

Figure 7.4. Measured and simulated offtake discharges for calibration

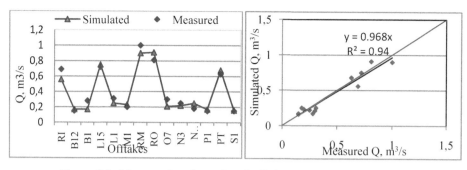

Figure 7.5. Measured and simulated offtake discharges for validation

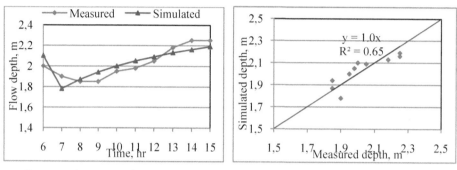

Figure 7.6. Measured and simulated flow depths at section 1+300 for calibration

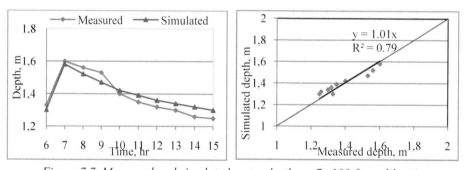

Figure 7.7. Measured and simulated water depths at 7+100 for calibration

Table 7.1. Calibrated parameters Chezy roughness coefficient C and discharge coefficients C_d

Chezy roughness coefficient (C)			Discharge coefficient (C_d)		
Lined canal (Main)	Unlined canals		Underflow		Overflow
	Main	Lat/sub-lateral	Free	Submerged	
65	34	30	0.65	0.55	0.95

Table 7.2. Statistical indicators of model performance for calibration and validation

Model statistics	Offtake flow		Model statistics	Flow depth (calibration)	
	Calibration	Validation		Section 1+300	Section 7+100
RMSE, m³/s	0.09	0.07	RMSE, m	0.075	0.042
NRMSE	0.21	0.17	NRMSE	0.04	0.03
CRM	0.06	0.03	CRM	-0.02	-0.07
ME	0.918	0.935	ME	0.998	0.998

7.5   Evaluation of the existing operation rules

7.5.1   Water diversion and demand

The simulated inflow diversion discharges under the existing operation are 10.3 m³/s for irrigation hours and 5 m³/s for off-irrigation hours. Hydraulic simulation under the existing operation rules has resulted in annual excess water diversion of about 27% or 41 Mm³. This excess diverted water is generally non-recoverable at Metahara Scheme, because of ever deteriorating groundwater quality in the area. A quick rise in the level of a brackish lake in the vicinity of the scheme has become a threat to the scheme, causing rising saline groundwater levels. Excess irrigation water mainly running to the tail reaches aggravates the rising groundwater levels, which are exposed to the surface at some spots. Substantial parts of the scheme area (North block of about 900 ha) have already become under a threat of high risk of salinization and waterlogging. The monthly diversion is lower than the demand only during the two months of July and August (Figure 7.8). During these months, the irrigation system is being partially closed for annual maintenance, with the assumption that rainfall would meet the water requirement. However, rainfall is too little, and unintended water stress occurs during these two months.

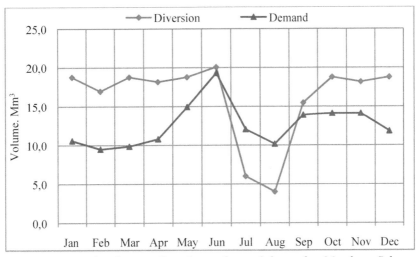

Figure 7.8. Simulated water diversion and actual demand at Metahara Scheme

The current operation rule hence has the following shortcomings in terms of operation of the main intake structure:

- *non-demand responsive operation.* In the current operation, little attention is given to monthly fluctuation of irrigation demand, and the head regulator is being operated in a more or less uniform manner over the period of 10 months. Although sugarcane is a perennial crop, monthly irrigation requirement varies owing to different planting dates and stages of the crop. There are two major drives for a more demand-responsive operation of the head regulator (main intake). First, to address the need for a more efficient water use in the basin as a result of increasing competition for water; if wasted water could be saved 2,000 ha of new land could be brought under irrigation. Second, it reduces tail runoff and hence waterlogging and salinization to ensure sustainability;
- *closure during July and August.* Irrigation is required throughout the year at Metahara Scheme unlike the current operation rule. Irrigation demand in July and August is very close to demands during the rest of the months. Even though the other reason for closure is annual maintenance, it could have been completed in one month (August), during which irrigation requirement is minimal;
- *incorrect gate settings.* In the existing operation, gate settings of the main diversion structure and assumed discharges are linearly related. However, this assumption underestimated the actual inflow discharge at the main intake. The discrepancy is particularly higher on the range of discharges over which the structure is commonly operated. A more realistic relationship based on simulated flows is given in Figure 7.9.

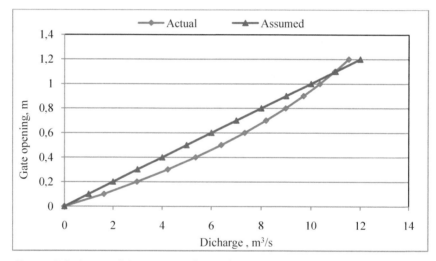

Figure 7.9. Assumed (existing) and actual relation between gate opening height and discharge at the main intake

7.5.2 Hydraulic aspects of elements of the canal network under the current operation

The relative size of the 11 km long main route of the network, which is exclusively unlined canal except the first 100 m length just downstream of the main intake structure along with all the offtakes, is shown in Figure 7.10. The concrete lined 100 m long head reach of the canal has a width of 10 m. At a distance of 5,070 m (5+070) from the head of the canal is a 10 m high chute (steep slopping drop structure). The reservoir RMC5, located at chainage of 6+500 to 7+100 has the layout as shown in the Figure by the wide section. There is a total water level difference of 24 m between the head and tail ends of the canal considered under the existing operation. Figure 7.10 also shows the simulated water levels in the parent canal, branches, laterals and sub-laterals.

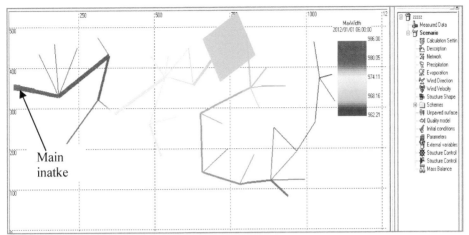

Figure 7.10. Canal sizes and simulated water levels of the canal system

Because of only 9 hours irrigation per day, the flow in the system is unsteady. Offtakes are closed during 15 hours of off-irrigation in a day during which all the sub-lateral (tertiary) canals have no flow (Figure 7.11). However, the off-irrigation flows in the branches and lateral canals are stored in the reservoirs as indicated by the colours in Figure 7.11. The simulated inflow during off-irrigation as per the prevailing current operation is 5 m³/s. For irrigation hours, the simulated discharges in different canals after the flow has attained a quasy steady state are shown in Figure 7.12, for a simulated inflow discharge of 10.3 m³/s.

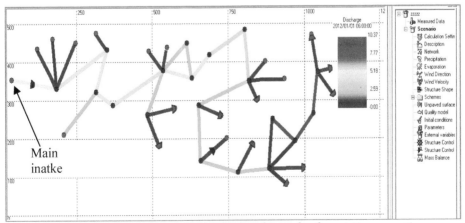

Figure 7.11. Discharges for off-irrigation hours for the existing operation

7.5.3   Hydraulic performance under the existing operation

Hydraulic performance of a water delivery system can be measured by indicators that show its effectiveness in water delivery. Flow rate (Q) and time (t) are factors determining water delivery volume and hence hydraulic performance (Figure 7.13). Intended volume of delivery is the amount of water based on the system design and the operator wishes to deliver at an offtake. Supplied volume is the amount of water actually delivered. All the supplied volume of water may not be effectively applied and stored in the root zone, and hence there are losses. Effective volume is a part of supplied water that is effectively controlled, applied, and used to replenish soil moisture deficit.

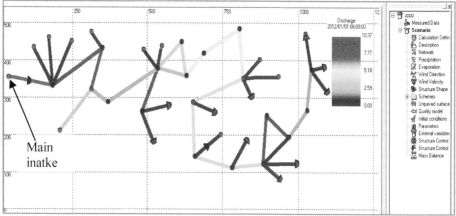

Figure 7.12. Discharges in different canals after a quasy steady state flow under the existing operation (end of irrigation hours)

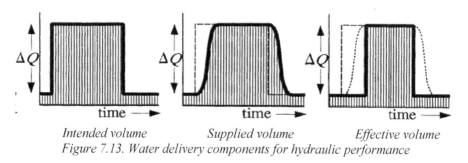

Intended volume   Supplied volume   Effective volume
Figure 7.13. Water delivery components for hydraulic performance

Performance ratio (relative delivery)

The performance ratio (relative delivery) was aggregated for head, middle and tail offtakes based on simulated flows (Figure 7.14). The simulated offtake flow rates over 9 hours (irrigation time per day) were considered as supplied volumes. Results indicate that tail reach offtakes were supplied with the largest amount of excess water delivery, while middle offtakes were in shortage. Head reach offtakes were on average supplied with nearly the intended delivery (excess of only 5%). The indicator tells only whether the intended amount of water has been delivered, but nothing about the proportion of supplied water used effectively. All offtakes are actually being kept open for 9 hours per day. Daily canal filling and emptying and associated head fluctuations are the main causes of the operations chaos. The reservoir (RMC5) on the canal at a chainage of 6+500 to 7+100 made differences in the hydrodynamic responses of the daily canal filling and emptying. In the current operation, the hydrodynamics respond such that the water levels (flow depth) in the main system upstream of RMC5 go on increasing from 6 AM to 3 PM (Figure 7.15). On the other hand, the water levels (depths) go on dropping downstream of the reservoir RMC5. Hydraulic flexibility (F) is a characteristic of a bifurcation which greatly affects water distribution. It is expressed as:

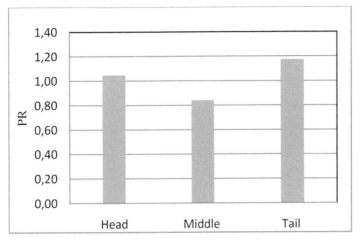

Figure 7.14. Performance ratio for head, middle and tail offtakes (DUFLOW)

$$F=\frac{S_{\text{offtake}}}{S_{\text{parent canal}}}=\frac{\Delta q/q}{\Delta Q/Q} \tag{7.7}$$

Where, F is hydraulic flexibility, S sensitivity, q is flow in the offtaking canal and Q is flow in the parent canal. Hydraulic flexibility also depends on the upstream and downstream water levels and is not a function of bifurcation point only. Hence, water level fluctuations have significant effects on flexibility.

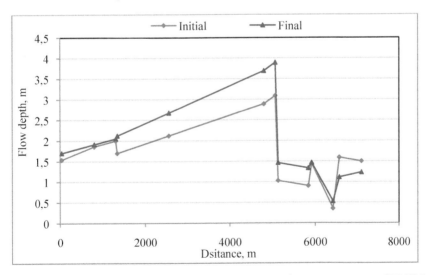

Figure 7.15. Flow depths along the canal at 6 AM and 3 PM upstream of RMC 5 (head reach)

The drop in head in the parent canal is higher in the middle reach (immediate downstream of the RMC5), and it becomes more or less constant at the tail reach (Figure 7.16). Typical discharge fluctuations for an offatke in the head, middle and tail reaches are shown in Figure 7.17. The higher drop in head on the immediate downstream of RMC5 is caused by operation of the reservoir outlet and hydraulic characteristics of the offtakes, and hence offtakes got a lower performance ratio. Under the current operation, reservoir storage highly depletes during the first three hours and causes a significant drop in head in the offtaking canal. The outlets are being opened fully and frequent adjustment is absent. The hydraulic characteristics of offtakes and water level regulators were not understood and not taken into consideration in the operation. Hydraulic flexibility (F) is the ratio of relative change in flow in the offtaking canal to that in the parent canal. Offtakes in the middle reach, except only one, are equipped with overflow structures and all water level regulators are underflow sluice gates. The sensitivity (S) of overflow structures is higher than that of underflow. Flexibility is hence higher than 1 for overflow offtake structures in the middle reach. Due to a more or less proportional division in the head reach, perturbations are more absorbed at the middle reach offtakes, letting water downstream. This eventually supplies tail offtakes with excess.

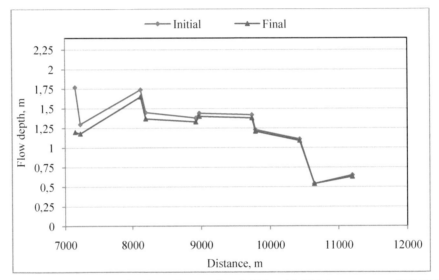

Figure 7.16. Flow depths along the canal at 6 AM and 3 PM downstream of RMC 5 (middle and tail reach)

Operational efficiency

Operational efficiency in this case applies to the efficiency of an offtake, and takes water delivery parameters like head, flow rate and timing into consideration. This is because operational losses which affect the effective volume are inevitable. Metahara Scheme, which is based on a daily canal filling and emptying, experiences significant operational losses particularly to drains and downstream ends. Here, operational losses in tertiary offtakes were considered, and losses in drains and at tail ends will be considered later in this Chapter. The effective volume of water delivery was determined for each offtake as shown in Table 7.3. The following criteria apply to the determination of effective volume (V_e) at a tertiary offtake (sub-laterals) and in lateral offtakes:
- field observation has revealed that the flow into a tertiary offtake during the first 30 minutes after offatakes have been opened will not be effective due to canal filling, low flow rate and head;
- when offtakes have been closed, the receding flow will be ineffective, which was determined from field observations to be about a volume of water delivered in 1

hour at a discharge rate of $0.5* Q_f$, where Q_f is the flow rate just before the offtake is closed.

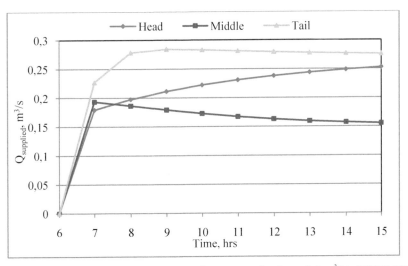

Figure 7.17. Typical offtake discharge fluctuation for $Q_D = 0.2 m^3/s$) (DUFLOW)

Table 7.3. Operational efficiency at lateral and sub-lateral levels

Offtake	V_s, m³	V_{inB}, m³	V_{inE}, m³	V_{inT}, m³	V_e, m³	OE
RI	20,700	949	639	1,590	19,100	0.92
B12	6,860	161	229	390	6,470	0.94
B1	6,860	161	229	390	6,470	0.94
C26	6,830	161	226	387	6,440	0.94
L15	19,200	498	600	1,100	18,100	0.94
L1	7,630	204	236	440	7,190	0.85
M1	6,050	162	187	349	5,700	0.94
O7	5,240	174	140	314	4,930	0.94
RO	29,700	1,540	786	2,330	27,400	0.92
N3	5,660	181	122	303	5,350	0.95
N8	5,320	83	143	226	5,090	0.96
N13	5,110	83	142	225	4,880	0.96
P1	6,710	148	196	344	6,370	0.95
RVI7	22,600	1,010	627	1,640	20,900	0.86
RV3	23,500	1,010	671	1,680	21,900	0.83
S1	8,370	205	248	453	7,920	0.77

The effective volume of water delivered depends also on whether the intended volume is fully met or not. Effective volume is always less than or equal to the intended volume. So, when the effective volume determined based on the aforementioned criteria is less than the intended volume, then that was considered the real effective volume. On the other hand, when the former is greater than the latter, then effective volume is taken to be equal to the intended volume. The reach-wise operational performance at lateral and sub-lateral levels is given in Figure 7.18. Efficiency at field levels is lower for the tail reach offtakes due to the fact that water level fluctuations in the parent canal in this reach are minimal. Still the operational efficiency at field levels is quite acceptable for a good management in each reach. On average, water loss at tertiary and field levels is only 9% of the total water delivery. For a surface irrigation system, this is fairly acceptable. However, there is a spatial variability from the head to tail ends.

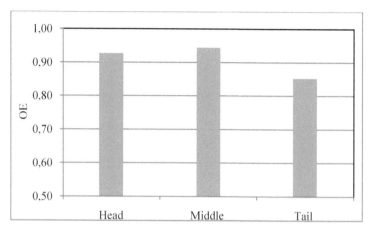

Figure 7.18. Operational efficiency at head, middle and tail offtakes (DUFLOW)

Equity (spatial uniformity)

Under the existing operation, the equity indicator was determined from the results of the hydraulic simulation. The spatial coefficient of variation (CV_R) was used as an indicator of equity for offtakes along the canal of 11 km long. The equity indicator for the head, middle and tail reaches and also overall equity were evaluated. The operation rules for the offatkes remain more or less the same over 10 months of the year, with little variation. As such, the equity was based on flows simulations of offtakes during irrigation hours of a day (9 hours).

The spatial coefficient of variation (CV_R), as an indicator of equity determined using both simulated supplied and effective volume of delivery at each offtake is given in Table 7.4. For each case, the equity levels within each reach ranked as 'good' according to the classification by Molden and Gates (1990). However, the overall coefficient of variation based on the supplied volume (CV_{Rs}) for equity indicator ranked as 'fair', while overall equity based on effective volume (CV_{Re}) was still 'good'. It is so, because although excess is being supplied to tail offtakes, the water is not effective due to operational losses or field percolation losses. This makes the effective volumes at the tail ends to be closer to the intended volumes, thereby increasing the overall equity indicator.

Table.7.4. Equity indicator based on effective and supplied volumes

Indicator	Head	Middle	Tail	Overall
CV_{Re}	0.05	0.05	0.01	0.10
CV_{Rs}	0.07	0.06	0.09	0.15

Water losses under the existing operation

Due to high salinity of groundwater and drainage water at Metahara Scheme, waste water is hardly recoverable. It was found out that, of the total annual excess diversion of 41 Mm^3 for Metahara Scheme, the proportion lost as percolation in the tertiary and feeder canals and on the field is only about 7%. Under the current operation, the simulation has proved that about 50% of the excess diversion is discharged at the tail end into shallow saline groundwater and waterlogged swampy fields. Sudden closure of tail offtakes and hydraulic characteristics of structures contribute a lot to this. Seepage at main level (main canals, branches and secondary canals) is relatively low, because of reduced infiltration rate of these canals due to clogging by the fine river materials. Seepage at the off-farm levels (distribution) rather accounts for a higher volume. The other significant contributor of water wastage is leakage and operational losses at the main levels. The proportion of water loss is shown in Figure 7.19.

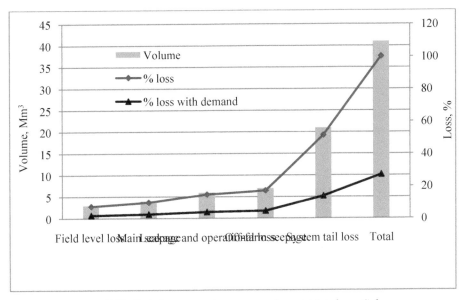

Figure 7.19. Existing water loss proportion at Metahara Scheme

7.5.4 Summary of failures of the existing operation with respect to hydraulic performance

The existing operation based on intermittent flow at tertiary levels (9 hours field irrigation) and unsteady continuous flow at the main levels posed considerable hydraulic performance problems. The performance ratio and operational efficiency at tertiary (sub-lateral) and lateral levels within a specific branch (secondary canal) performs well as was shown in the previous sections. From water saving point of view, the overall equity in terms of the ratio of supplied to intended volumes is more important. Hence, the overall spatial coefficient of variation, CV_R of 0.15 for the canal system considered indicates a low level of equity. The equity determined for sub-laterals fed from different branch canals is even worse as was determined in Chapter 6 using flow monitoring. In this case in addition to the offtakes on the simulated canal (Branch 1 in Figure 7.24), the flows were monitored at 10 offtakes, 5 on Branch 1-1 and 5 on branch 2-2 over three months. The equity levels for offtakes on the simulated branch and other two branches are shown in Table 7.5. A spatial overall equity (CV_R) of 0.21 shows an inadequate equity for a good water distribution.

Table 7.5. Equity of water deliveries for offtakes on different branches

Branch 1 (simulated)	Branch 1-1	Branch 2-2	Overall
0.15	0.13	0.19	0.21

The inadequate hydraulic performance of the existing operation is associated with the following features:

- *flow unsteadiness*. The highly unsteady flow condition of the existing operation is the cause for significant operational losses at the main level, tail runoff, and unsteady offtake delivery;
- *inequity within branch canal levels and across branches*. For the canal system simulated, although the equity levels within each reach are adequate, inequity is a major operational failure within a branch canal and across;
- *leakage and operational losses*. Significant leakage and operational losses commonly occur under the existing operation at division points and control structures. High day time discharges at main levels and the need to alter the flow rates to meet requirements causes leakage into drainage systems and ultimately to poor quality water courses;
- *tail runoff losses*. Waterlogging as a result of excess runoff and percolation losses at the tail ends is a major water management problem under the existing operation. In fact, salinization has been a major threat to the sustainability of the scheme as nearly about 900 ha of land has already been under threat.

7.6   Simulation of operation scenarios for improved hydraulic performance

The major shortcomings of the existing operation rules concern operation of the regulator of the main intake, night storage reservoirs, and offtakes. The following three scenarios were simulated with DUFLOW to match diversion and demand, reduce tail excess runoff and ensure better hydraulic performance;

- *24 hours field irrigation and constant inflow at the main intake*. Inflow at the main diversion may be kept constant over 24 hours in this case. The existing 9 hours irrigation per day necessitates a higher and steady stage in the canal in order for the offtakes deliver their design discharges. However, it is hardly possible due to parent canal emptying and filling;
- *12 hours irrigation and modified night storage reservoir operation*. Increasing the irrigation time to 12 hours per day and modifying the operation of the reservoir so as to maintain fairly steady water levels in the middle reach;
- *Adopting the existing 9 hours irrigation*. While adopting the current field irrigation duration, operational measures at the main intake, reservoirs and timing of closure for offtakes will be considered.

7.6.1   Scenario 1: 24 hours irrigation

Diversion versus demand (water saving)

The annual net irrigation demand for Metahara Scheme is 151 Mm^3. Taking into consideration the main and off-farm seepage and losses at tertiary levels (about 10%)

and leaching of 5%, a total loss of 15% of the demand was considered in this operation. A constant inflow diversion of 6.3 m³/s over ten dry months (September-June), and an inflow in July and August of 4.6 and 3.9 m³/s respectively will meet the field demand. The unavoidable losses will also be satisfied with this flow rate. This would result in annual water saving of 15% of the demand, which is equivalent to about 18 Mm³. This corresponds to a volume of water required to irrigate about 1,100 ha of land on the downstream of Metahara Scheme. This fresh water saving means a lot in this semi-pastoral semi-arid region, where water is increasingly scarce in respect of large irrigation expansion. Figures 7.20 and 7.21 show the canal profile and the steady discharges and flow depths along the main canal route for this scenario of operation.

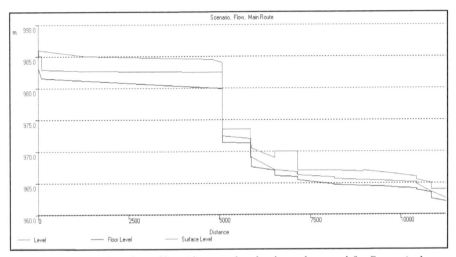

Figure 7.20. Canal profile with water levels along the canal for Scenario 1

Relative offtake delivery (performance ratio)

A canal infiltration rate of 1 mm/hour has been assumed for seepage losses in the main and off-farm system. Discharge rates equivalent to this seepage were considered in the simulation of flows in the scenarios. Simulated steady state offtake discharges and intended (required) flow rates over 24 hours were determined for each offtake along the canal as shown in Figure 7.22. Steady state flow in the canals will call for a different operational setting of flow control structures than the current setting. It requires mainly full opening of the structures. In this scenario, for underflow sub-lateral and lateral offtakes the flow openings can be increased uniformly by 10 and 20 cm respectively to ease the operation. The increase in the flow openings is needed because of low stage in the parent canal due to a reduced and steady flow in the canal. Similarly, for overflow offtake structures, uniformly lowering the crest levels of movable crest weirs (Romijn weirs) by 10 cm is required to ensure steady offtake flow. The reach-aggregated relative delivery (performance ratio) as a spatial average of the ratio of supplied to intended discharge rates is adequate and right. In each reach, the performance ratio (Dubois and Prade) is nearly unity except that it is a little bit higher for the tail each (Figure 7.23).

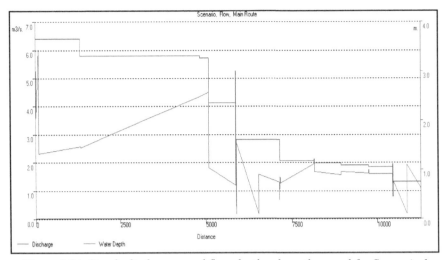

Figure 7.21. Steady discharges and flow depths along the canal for Scenario 1

Equity (spatial uniformity)

The equity of offtake water delivery was determined based on the simulated steady state and intended discharges for continuous flow at the offtake. In this scenario, equity is best achieved as the main cause of spatial non-uniformity of delivery is unsteadiness. This operational scenario, based on 10 and 20 cm change in the operational settings of the structures for sub-lateral and lateral offtakes respectively throughout, avoids significant temporal fluctuations in offtake discharges and offtake operational losses. This means that it increases the efficiency of water use and reduces losses due to flow advance and recession within tertiary units. The spatial coefficient of variation (CV_R) of the ratio of simulated (supplied) to intended discharges is 6%, which makes a delivery system with a 'good' level of equity.

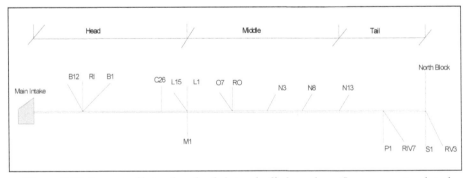

Figure 7.22. Layout of lateral and sub-lateral offtakes whose flows were simulated

Reduction in tail runoff

In the existing 9 hours irrigation tail run-off losses account for about 14% of annual demand and 51% of total water losses. It accounts for 21 Mm^3 of annual water losses into saline shallow groundwater. In the modified 24 hours irrigation and steady state water levels and discharges in the main system and at offtakes tail run-off will be substantially reduced. The tail losses are avoided in this operation for two main reasons. First, absence of closure of tail reach offtakes produces no operational loses and hence no significant water will be released at the tail end into the drainage (escapes. Second, the hydrodynamic behaviour does not deliver a significant over supply as in the existing operation. Of this tail run-off about 13 Mm^3, is discharged at the tail end of an irrigation block called North Block (900 ha), which is being supplied by a lateral from the tail of the canal considered in this simulation. The irrigation canal considered for simulation consists of the main canal and branch 1, with a combined length of 11 km (Figure 7.24). The North Block (lateral), RV3 and RVI7 are laterals branching from the tail reach of the canal system simulated. The remaining 14 offtakes simulated are on the upstream of the Offtake for RV$_3$. While nearly 4 Mm^3 is a tail run-off from the sub-laterals RIV$_7$ and RV$_3$, the remaining 4 Mm^3 is from other branch canals than considered in this simulation (branches 1-1, 1-2, and 1-3 in Figure 7.24).

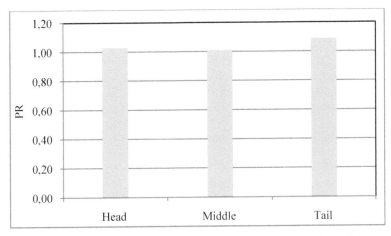

Figure 7.23. Performance ratio for Scenario 1 in each reach

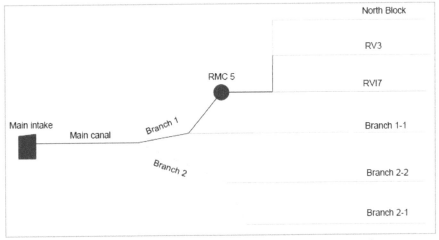

Figure 7.24. Main system and branches layout of Metahara Scheme
(Simulated canal in blue)

Table 7.6. Tail loss reductions for Scenario 1

	Existing annual tail run-off, Mm^3	Scenario 1, Annual tail run-off, Mm^3
North Block	13	5
Laterals to RIV$_7$ and RV$_3$	4	-
Other branch canals	4	-
Total	21	5

7.6.2 Scenario 2: 12 hours irrigation

Increasing the field irrigation time by 3 hours would result in an increase in the overall water delivery/hydraulic performance. In this scenario, water diversion at the main intake takes place 24 hours a day; however, field irrigation occurs 12 hours a day.

Water diversion versus demand

In this case the inflow at the main diversion is kept constant over the irrigation and off-irrigation hours. This ensures relatively steady water levels in the parent canals. Considering seepage losses in the main and off-farm (10% of demand) and leaching of 5%, a diversion inflow rate of 8 m^3/s for 10 months and of 4 m^3/s for the wet months of July and August during irrigation hours is considered. On the other hand, a constant diversion rate of 4 m^3/s, to be stored in the reservoirs during off-irrigation hours (night) is considered to meet the total demand and losses (Figure 7.25). This option has a lower water saving potential than Scenario 1. The annual tail runoff in this case amounts 9 Mm^3, while it is 5 Mm^3 for Scenario 1. The annual saving in the volume of diverted water is 16 Mm^3 compared to the existing operation or 11% of the demand.

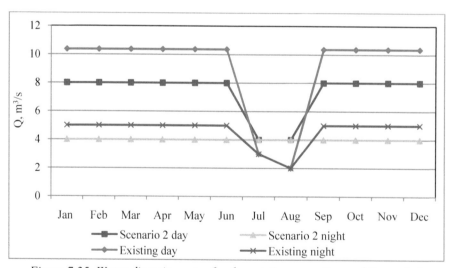

Figure 7.25. Water diversion rates for the existing operation and Scenario 2
(Annual water saving = 16 Mm^3)

Relative delivery (performance ratio)

The relative delivery was determined as a ratio of volume of water supplied to offtakes to volume of water intended. Unlike Scenario 1, in this case the offtake flows are unsteady and volumes of water delivered in 12 hours were considered instead of discharges. Table 7.7 shows the offtakes considered and the intended discharge rates for 12 hours of irrigation. The intended discharges were determined based on the intended discharge rates of the existing operation (9 hours irrigation). The performance ratio for this operation scenario shows superior performance for each reach (Figure 7.26) and for the whole canal system.

    This scenario can be achieved with uniform 5 and 10 cm changes (increase) in the gate settings of sub-lateral and lateral offtake structures from the existing operation, and with an extended offtake flow time. This happens because of the modified (reduced) inflow rate at the main intake, which also causes a corresponding drop in water levels in the parent canal. It can be observed from Figure 7.27 that the discharges of offtakes at the head, middle and tail reaches remain almost constant and nearly at the intended flow rates. Moreover, this operation avoids a significant drop in head in the middle reach of the parent canal. Figure 7.28 shows the discharges along the canal in the morning when offtakes are opened and in the evening when the offtakes are closed for this Scenario. The discharge at offtake opening shows the off-irrigation flow in the canal where all tertiary offtakes are closed. The drops in the discharge along the canal during off-irrigation hours are flows to branch (lateral) canals for night storage.

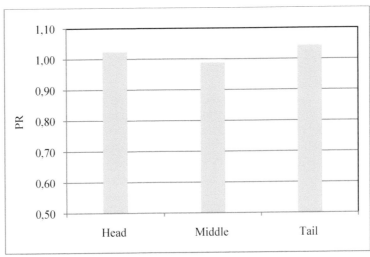

Figure 7.26. Performance ratio for Scenario 2

Table 7.7. Offtakes along the canal and their intended flow rates for 12 hours of irrigation

	Head reach					Middle reach						Tail reach			
Offtake	RI	B12	B1	C26	L15	L1	M1	O7	RO	N3	N8	N13	P1	RIV7	RV3	S1
Q_i, m³/s	0.5	0.15	0.15	0.15	0.5	0.15	0.15	0.15	0.8	0.15	0.15	0.15	0.15	0.5	0.5	0.15

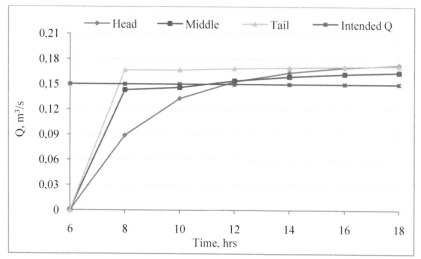

Figure 7.27. Typical offtake discharge fluctuation under Scenario 2 (for Q_i = 0.15 m^3/s)

Operational efficiency

Operational losses at lateral and sub-lateral levels are inevitable in this scenario of
operation due to initial low stage, low flow rate and closure of offtakes. The effective
(V_e) volume of supply was determined for each offtake. As there is no or low flow in
the offtake canals at the time of opening, the flow during the first 30 minutes is
ineffective due to canal filling. Moreover, the receding flow after offtake closure is
ineffective as demonstrated by field observations. This ineffective volume is estimated
to be the flow over 30 minutes at a rate of half of the offtake flow rate at closure. The
offtake and aggregated operational efficiency for each reach is given in Table 7.8. The
operational efficiency in each reach for this operation scenario is 'good' according to
Molden and gates (1990).

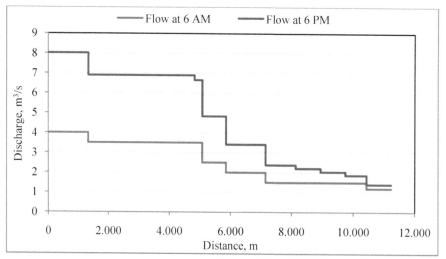

Figure 7.28. Discharges along the canal at offtake opening and closing for Scenario 2

Equity

Equity can be determined using the supplied volume or the effective volume. The objective of operation for equity is actually to achieve equity of the effective water delivered and not just the supplied water. Effective volume, the volume of water applied to the field and stored in the root zone to replenish the soil water deficit was determined as indicated in Table 7.8. The intended volume is based on 12 hours irrigation at flow rate indicated in Table 7.7. The spatial coefficients of variation of the ratio of effective to intended and supplied to intended volumes are given in Table 7.9. The indicator shows that the equity levels are adequate for each reach and overall (CV_R less than 0.1) with a small exception at the head reach for which $CV_{Rs} = 0.11$.

Tail runoff reduction

The tail run-off for the 12 hours irrigation is lower than the existing operation; because the flow in the system at the tail is lower, which could be easily controlled. This reduces the operational losses at closure of tail offtakes. Tail run-off in this case is 12 Mm^3 compared to 21 Mm^3.

Table 7.8. Operational performance for operation scenario 2

Offtake	V_s, m^3	V_{inB}, m^3	V_{inE}, m^3	V_{inT}, m^3	V_e, m^3	OE	Reach
RI	25,800	1,450	1,180	2,630	23,200	0.90	Head
B12	5,840	131	302	433	5,410	0.93	0.92
B1	5,840	131	302	433	5,410	0.93	
C26	6,250	114	331	444	5,810	0.93	
L15	23,500	1,450	1,030	2,480	21,000	0.89	
L1	6,660	253	303	556	6,110	0.92	
M1	7,040	268	320	588	6,460	0.92	
O7	6,450	278	288	566	5,880	0.91	Middle
RO	32,100	2,080	1,390	3,460	28,600	0.89	0.91
N3	6,390	270	295	565	5,830	0.91	
N8	6,600	277	299	576	6,020	0.91	
N13	6,560	277	294	571	5,990	0.91	
P1	6,980	301	308	609	6,370	0.91	Tail
RIV7	23,000	1,610	987	2,590	20,400	0.89	0.90
RV3	22,500	1,670	956	2,620	19,920	0.88	
S1	6,380	276	282	558	5,826	0.91	

V_{inB}= ineffective volume at the beginning of inflow, V_{inE}= ineffective volume at the end of inflow, V_{inT}= total ineffective volume

Table 7.9. Equity indicators for operation scenario 2

	Head	Middle	Tail	Overall
CVRe	0.09	0.05	0.04	0.07
CVRs	0.11	0.04	0.04	0.08

7.6.3   Scenario 3: 9 hours irrigation with modification in operational setting

The operational modifications considered in scenario 3 include the main intake, reservoir (RMC5) outlet and offtakes downstream of the reservoir.

Water diversion

A day hours diversion rate of 9 m³/s and a night hours diversion of 4 m³/s over 10 months (September to June) are assumed in this scenario. Over the rainy months of July and August, diversion rates of 4 and 3 m³/s are assumed for the day and night hours to meet the field demand and unavoidable losses. The annual volume of water diverted will be about 172 Mm³. Compared with the existing diversion, the annual reduction in the volume of diverted water is about 14% of the annual demand, and is equivalent to about 21 Mm³. Figure 7.29 shows the water levels and discharges along the canal just before offtakes closure. Although water levels increase and decrease gradually in time upstream and downstream of RMC5 respectively, the fluctuation is minimum compared to the existing operation.

Performance ratio

The operation of the main intake, reservoir RMC5 and offtakes would have to produce different hydrodynamics than the existing, while duration of field irrigation remains the same. The existing operational setting of offtakes, but with modified main intake and RMC5 reservoir outlets can be employed for this scenario. The main intake structure has to be operated with a reduced outlet opening (0.8 m opening height instead of the existing 1 m opening). Similarly, instead of full opening of the outlet of RMC5, it is reduced by 0.6 m (total opening of 1.2 m) during irrigation hours, because of the reduced discharge. This also helps to keep fairly steady water levels in the parent canal downstream of the reservoir. Offtakes upstream of RMC5 can be operated in the same condition as the existing. However, downstream of the reservoir, an increase in gate openings of sluice gate offtakes by 15 cm and lowering the crest levels of overflow offtakes by the same amount is assumed. The reach PR for this operation scenario is shown in Figure 7.30. The PR in the middle reach has an inferior level. Still it is acceptable as it ranks 'fair' in the performance standard.

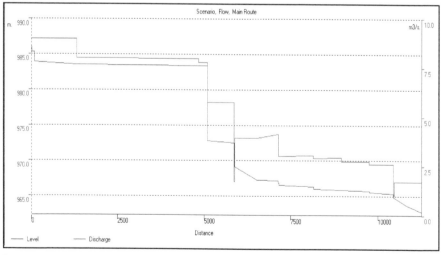

Figure 7.29. Quasy-steady water levels and discharges along the canal for Scenario 3 at the end of irrigation hours

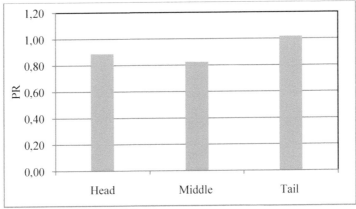

Figure 7.30. Performance ratio for Scenario 3

Operational efficiency

As this also involves intermittent flow at offtakes, operational losses do occur at offtake openings and closures. The water delivery components and operational efficiency (OE) for this scenario are shown in Table 7.10. As shown, the aggregated operational efficiency for each reach is good enough for an acceptable equity. Both for scenario 2 and 3 as well as for the existing operation, the water losses downstream of tertiary offtakes (within tertiary units) are very low, and hence offtake operational efficiency in each case is acceptable.

Table 7.10. Operational efficiency at lateral and sub-lateral levels for scenario 3

Offtake	V_s, m^3	V_{inB}, m^3	V_{inE}, m^3	V_{inT}, m^3	V_e, m^3	OE	Reach
RI	18,300	1,900	1,050	2,950	15,400	0.84	
B12	5,480	324	173	497	4,980	0.91	Head
B1	5,480	324	173	497	4,980	0.91	0.90
C26	5,540	328	175	503	5,040	0.91	
L15	16,600	975	521	1,500	15,142	0.91	
L1	6,760	401	211	612	6,150	0.91	
M1	5,360	320	167	487	4,870	0.91	
O7	5,440	350	161	511	4,920	0.91	Middle
RO	28,000	2,910	766	3,680	24,400	0.87	0.91
N3	5,570	309	157	466	5,110	0.92	
N8	5,240	158	160	319	4,920	0.94	
N13	4,890	156	152	308	4,580	0.94	
P1	6,140	290	189	479	5,660	0.92	Tail
RVI7	21,600	2,010	606	2,610	19,000	0.88	0.90
RV3	21,400	1,930	617	2,550	18,900	0.88	
S1	5,880	305	183	488	5,390	0.92	

Equity

The equity level for this operation scenario was also determined based on both effective and supplied volumes in relation to the intended volume. The spatial coefficients of variation (Table 7.11) indicate that the reach and overall equity levels all perform well.

Tail runoff

In this scenario, it is shown that in the existing operation, the huge downstream runoff is exclusively a result of inadequate operation. For the canal considered, it was shown that operation of the reservoir plays the most important role in water distribution on the downstream and on tail losses. While fairly meeting the hydraulic performance standards, this operation avoids unnecessary high outflow rates during the initial 2-3 hours after the reservoir has been opened and depletion afterwards. As has been discussed in Chapter 6, the hyper-proportional nature of offtakes in the middle reach would have produced significant flow accumulation and drainage at the tail as in the existing case. The combined modified operation of the main intake, RMC5 and middle and tail offtakes would reduce the tail runoff to about 7 Mm^3.

Table 7.11. Equity (*CV*) for Scenario 3

	Head	Middle	Tail	Overall
CVRe	0.08	0.04	0.08	0.10
CVRs	0.09	0.06	0.10	0.12

Conclusion

Three operational scenarios that would enhance water saving and sustainability in Metahara Scheme were considered and simulated. These operation measures, while each has adequate hydraulic performance, also have significant contribution for water saving and for control of rising saline groundwater levels in the scheme. As water losses are inevitable in such a gravity irrigation scheme, unavoidable losses such as main and off-farm canal seepage, leakage at branching and division points, operational losses at main system level, and leaching requirements have been considered in the proposed operational scenarios. Annual water diversion, tail runoff and water saving are summarized in Table 7.12. It might seem that the amount of water saved (11-14%) is minor; however, it is not even easy to achieve. Hence, these water savings if could be achieved are significant taking into account the historical water management situation in the scheme.

Table 7.12. Annual water diversion and saving for operation scenarios

	Existing operation	Scenario 1	Scenario 2	Scenario 3
Annual water diversion, Mm^3	193	175	177	172
Annual tail run-off, Mm^3	21	5	9	7
Annual water saving, Mm^3	-	18	16	21
Annual water saving, % demand		12	11	14

8 PERFORMANCE ASSESSMENT IN COMMUNITY MANAGED SCHEMES USING COMPARATIVE INDICATORS

8.1 Data for comparative performance evaluation

Data required for comparative performance assessment and the methodology used for the acquisition are summarized as follows.

Field survey

A comprehensive field survey was made to each irrigation scheme by a walk through the different components of the schemes. The objectives were:
- to quickly get acquainted with the sources of irrigation water and conditions of the water supply;
- to physically assess and evaluate the water diversion head works;
- to understand the water conveyance and distribution systems and quickly evaluate their conditions;
- to understand the existing irrigation scheduling and operation of flow control structures;
- to assess on-farm and off-farm irrigation water management practices.

Moreover, the field survey enabled measurement of some components such as dimensions of intakes, main canal sizes and tertiary offtakes. Field survey is of course an unavoidable activity in performance evaluation as it provides lots of information in a relatively short period of time.

Landholding of farmers

In Ethiopia, about 65% of farming households operate on land sizes of less than 1 ha; while about 40% rely on a land size of 0.5 hectare or less (Gebreselassie, 2006). The average size of landholding of smallholder households in Ethiopia is 0.7 ha. In fact, landholding size is one of the major factors that constrain farm income and the level of household food security particularly in the Ethiopian highlands. For Golgota and Wedecha schemes, average landholdings of farmers at head, middle and tail reaches of each scheme was determined using a questionnaire and is given in Table 8.1. Size of landholding at Wedecha Scheme is relatively small and it is one of the major limitations for increasing household income and improving livelihoods in the area.

Questionnaire survey

In the community managed schemes, relevant data for comparative performance, such as agricultural output, landholding, cropping pattern and intensity, amount of irrigation water delivered at various locations within the schemes, etc, are hardly available. To this end, a questionnaire survey to the water users themselves was made to collect primary data related to irrigation service, agricultural produce, landholding size, livelihood from irrigated agriculture, etc. The survey was made at each scheme using sampled water users from the head, middle and tail reaches of the schemes.

Table 8.1. Landholding characteristics of Golgota and Wedecha schemes

Scheme	Sub-system	Average landholding, ha			
		Head reach	Middle reach	Tail reach	Average
Golgota		0.9	1.6	1.0	1.2
Wedecha	*Godino*	0.7	1.0	1.0	0.9
	Gohaworki	0.3	0.3	0.4	0.33

Flow measurement (Parshall flumes and stage-discharge relation)

Measured irrigation flow data were not available and is given less priority at the community managed schemes under consideration. However, irrigation flow measurement is among the key data for comparative irrigation performance assessment. So diverted irrigation flow measurements were made for 2010/2011 and 2011/2012 agricultural year (September 2010 to August 2011 and September 2011 to August 2012) at each scheme. For Wedecha Scheme (Godino and Gohaworki sub-systems), flow measurements were made using Parshall flumes, with level readings made three times a day. However, for Golgota Scheme, an alternative method was used due to larger canal sizes. A staff gauge was used to measure water depths in the canal at different discharges being measured with current meters. A stage-discharge $(Q$-$h)$ relation was used to determine flows for any other observed stage. The basic rating curve equation $(Q$-h relation) for open channel flow used is:

$$Q = K * (h-h_o)^m \qquad (8.1)$$

Where, Q is discharge (m^3/s), h is stage in the canal (m), h_o is stage at which flow is zero (m) and k and m are constants. The coefficients k and m were determined from a linear plot of log Q versus log h by a linear regression.

Irrigation water was measured at two locations in the canal. The volume of diverted (supplied) irrigation water from the source was measured at the head of the main canal. On the other hand, the volume of delivered irrigation water was measured at the head of the command. Irrigation water being a major input to agriculture, data on irrigation flow were used to evaluate both indicators of water supply and water productivity, which are key for comparative performance assessment.

Irrigable and annual irrigated area

Irrigable land could either be fully or partly utilized for cropping throughout the year depending on various factors. Irrigable land, the land which could nominally be irrigated with the designed irrigation infrastructure for the schemes was available at local agricultural development offices. It was also determined by surveying the areas with the global positioning system (GPS) for each scheme. Annual irrigated area is the sum of the areas under irrigated crops during all cropping seasons in a year, and depends on irrigation intensity. While data on irrigated land are available at local agricultural development offices, they were supplemented using the questionnaire survey (irrigated land holding of sampled farmers, irrigation intensity and total number of farmers).

Agricultural production

Irrigation water management is ultimately meant to enhance agricultural production through sustainable water use. Secondary data on agricultural production are commonly ambiguous for research purposes and these data were better collected from primary sources, specifically from the schemes under consideration. As such, data on the yield (agricultural output) for 2007/08, 2008/09 and 2009/10 were collected from the farmers at each scheme during October 2010 through March 2011. Data on agricultural produce was collected together with data on landholding from a sample of water users at different reaches. From the average landholding and total number of irrigators, total annual production and value of production was determined. Secondary data on agricultural produce from district agricultural development bureaus was also collected for comparison.

8.2   Data analysis and interpretation

8.2.1   Water supply indicators

Annual relative water supply (ARWS) and annual relative irrigation supply (ARIS) were evaluated for the agricultural years of 2010/11 and 2011/12 (September-August) for each irrigation scheme. Average values of water/irrigation supply over the two agricultural years were considered. Annual values of four water supply/demand values were determined: namely annual water supply, annual crop water demand, annual irrigation supply and annual irrigation demand. Annual irrigation supply is the volume of irrigation water delivered to the head of the command. Annual water supply is the sum of delivered irrigation water and effective rainfall. Annual crop water demand is the actual evapotranspiration demand of the crops, determined using CROPWAT model with the FAO guidelines (Allen et al., 1998) for a given cropping pattern and irrigation intensity. Irrigation demand is crop water demand less effective rainfall. Water supply indicators for Golgota Scheme and the two sub-systems of Wedecha Scheme are given in Table 8.2 and Figure 8.2. Water supply per unit irrigated area (Table 8.2) at Golgota Scheme is about three times as high as the average at Wedecha Scheme. Institutional arrangements for decision making on water management are the major factor for a higher unit irrigation supply at Golgota Scheme.

Table 8.2. Annual irrigation water supply per unit irrigated area

Scheme	Sub system	Irrigated area, ha	Annual irrigation water supply per unit irrigated area, m³/ha	Annual irrigation supply, mm
Golgota		550	31,300	3,130
Wedecha	Godino	200	8,020	802
	Gohaworki	50	11,500	1,150

From Figure 8.2, it can be observed that ARIS values are greater than ARWS values for each scheme, which indicates irrigation is the major source of water supply for agriculture in the areas. It can also be observed that the ARIS values for each scheme are higher than 1.0, depicting that, disregarding the distribution of the supply over the months, excess irrigation water was supplied. It is interesting to note that more than three folds of annual irrigation demand is being supplied for Golgota Scheme (ARIS = 3.17), followed by nearly twice of irrigation demand for Gohaworki Sub-system of Wedecha Scheme (ARIS = 1.90). Irrigation efficiency in these schemes is extremely low due to poorly constructed and maintained water distribution canals and poor off-farm water management. On-farm irrigation practices are also characterized by poor water control and inequity of distribution (Figure 8.1). In addition to the diverted flow, measurement of irrigation water delivered to individual plots enabled determination of the combined conveyance and distribution efficiency in each scheme. The combined conveyance and distribution efficiency, excluding the field application efficiency, for Golgota Scheme is 70%, while that for Wedecha Scheme (both sub-systems) is 60%.

a. *On-farm irrigation at Golgota Scheme*   b. *Poor water distribution at Wedecha Scheme*

Figure 8.1. Poor on-farm and off-farm water management at Golgota and Wedecha schemes

Excess irrigation water supply to Golgota Scheme is due to two important factors:

- *water diversion responsibility.* At this scheme, farmers themselves are responsible for the volume of water diverted from the river. The amount of water diverted to Golgota Scheme main canal is being decided by the management committee of the WUA of the scheme. As such, critical assessment of irrigation demand is absent and water is being diverted without due consideration of the demand schedule. So, in addition to annual excess, monthly variations of the Relative Irrigation Supply (Molden *et al.*, 1998) are high for the scheme;

- *absence of irrigation water fee.* Another important factor for excess irrigation water supply is the fact that there is no irrigation water fee at Golgota Scheme. Farmers of the scheme have been using water for free since ever, and as such there is no incentive for saving irrigation water.

Farmers at Wedecha Scheme (Godino and Gohaworki sub-systems) on the other hand pay an annual irrigation water fee of about 48 US$/ha to a regional government irrigation authority. This irrigation water fee is for the services rendered by the agency in operation of the main reservoir outlet (headwork) and some maintenance related to the headwork. Introduction of water fee at Golgota Scheme is a feasible intervention for both as an incentive for saving precious water and for reducing future risks of waterlogging and salinization due to excess irrigation.

For Wedecha Scheme, Godino and Gohaworki sub-systems are supplied with water from a reservoir from which water is released and conveyed over 5 km through the main river channel. Release of water from the reservoir is made based on requests from the water user associations. However, the water request for both sub-systems is released through the same channel from the reservoir. Significant variations do exist in water supply indicators even for the two sub-systems of the same scheme (Figure 8.2). Excess irrigation supply at the heads of the commands of Gohaworki Sub-system (ARIS = 1.90) compared to that of Godino Sub-system (ARIS = 1.20) was observed. The flow control sluice gates for regulating flows into the canals at each diversion for the sub-systems (Figure 8.3) have been demolished by farmers. They use locally available stones and wooden logs to raise water levels and facilitate diversion. Their main reason for demolishing the gates is lack of trust on the availability of enough water. The diversion structure of Gohaworki is located on the upstream giving it more advantage. Due to lack of decision making over the release of water from the reservoir and being located on the upstream, farmers at Gohaworki usually divert as much irrigation water as possible into their canal, regardless of their demands.

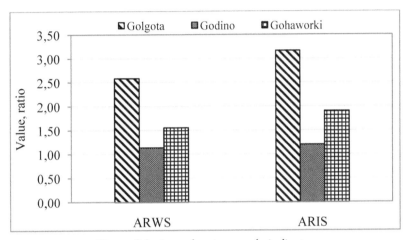

Figure 8.2. Annual water supply indicators

While participatory decision making on irrigation water management in the community managed schemes would mean a more efficient water use, effectiveness depends on certain factors. Sound irrigation scheduling, good physical infrastructure for water control, mechanism for flow monitoring or flow measurement, measures for illegal acts, etc. are crucial for saving water. In the absence of these conditions for appropriate off-farm water management, there occurs a large inequity between sub-systems or groups of water users as in the case of Wedecha Scheme.

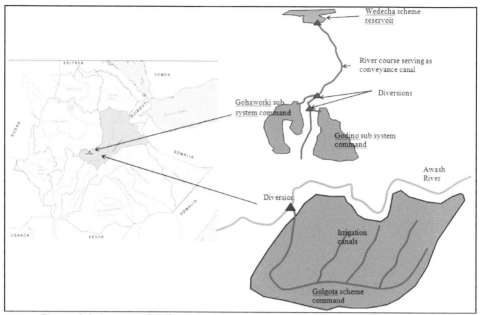

Figure 8.3. Layout of Golgota and Wedecha community managed schemes

8.2.2   Agricultural output indicators

Land productivity

The output per unit of irrigated area or command area does not necessarily imply the conditions of irrigation water supply. In fact there are other important factors affecting land productivity, such as soil type and agricultural inputs. Where land is a constraining factor for production, land productivity as indictor becomes important. Still, land productivity and water productivity are interrelated. In the community managed schemes, landholding size is a major constraint for production and improving household income. For the years 2007/2008, 2008/2009 and 2009/2010, total agricultural production and thus outputs from the produce at local market prices (US$) were determined for each scheme. The size of irrigated cropped area over the three years period remained same at each scheme. At Golgota Scheme, only one crop (onion) is grown three times a year while the other crops are grown twice a year. Cropping intensity at both sub-systems of Wedecha Scheme is 200%, each crop being grown twice a year. Irrigated and nominal (command) areas and annual outputs are given in Table 8.3.

Figure 8.4 shows that the values of indicators (agricultural outputs) for these schemes are superior compared to particularly Sub-Saharan Africa. With average of 873 US$/ha in 2006, land productivity in Sub-Saharan Africa was the lowest in the world. This was about 56% of the world average and 41% of East Asia (International Food Policy Research Institute ((IFPRI, 2008). In the schemes under consideration, annual output per unit area irrigated is particularly high as a result of a more intensive irrigation. In the schemes, highly marketable crops such as Onion and Tomato are the major crops, whose growing season is relatively short. These crops are harvested two or three times a year. While land productivity highly depends on cropping intensity, it is also dependent on water availability. Higher irrigation intensity is responsible for higher land productivity at Golgota Scheme. Irrigation water availability and management is a key element for increased irrigation intensity. At Golgota Scheme, farmers are all responsible for the volume and timing of irrigation water diverted, which made the water supply more dependable. This made the irrigation more intensive, which in turn increased land productivity at Golgota Scheme.

Table 8.3. Irrigated/nominal command areas and annual agricultural output indicators

Scheme	Sub-system	Annual harvested area, ha	Nominal command area, ha	Annual output per irrigated area, US$/ha	Annual output per command area US$/ha	Output per harvested area, US$/ha
Golgota		1,375	600	6,400	5,870	2,670
Wedecha	*Godino*	400	300	4,570	3,040	2,280
	Gohaworki	100	60	3,300	2,750	1,650

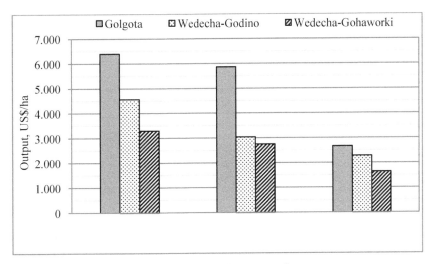

Figure 8.4. Land productivity indicators

Another factor for land productivity at these schemes is willingness of farmers to invest more on their piece of land. A questionnaire survey to water users and a focus group discussion has confirmed that willingness of farmers to invest on their plot of land greatly depends on the size of landholding. The larger the farm size of a farmer, the higher is his willingness to invest more on his plot. This was reflected in the higher productivity values of Golgota Scheme (Figure 8.4). Willingness by farmers to invest more is also related to their degree of confidence on water availability, which is again much better in the case of Golgota Scgeme. As a result, agricultural indicators perform much better for Golgota Scheme.

While irrigation intensity at both sub-systems of Wedecha scheme is same, there is a significant difference in the outputs particularly in the annual output per unit irrigated area (Figure 8.4). This can be explained well by the willingness to invest more. The average landholding size is 0.9 ha and 0.3 ha for Godino and Gohaworki sub-systems respectively. It was found that farmers at Godino sub-system are willing to invest more on their plot of land in terms of other agricultural inputs in addition to water, such as fertilizers, pesticides and herbicides, which means higher yield per unit of land. Lower land productivity at Gohaworki sub-system of Wedecha Scheme is a combined effect of less dependability of irrigation supply and lack of willingness to invest more due very small landholding size. With average landholding of 0.3 ha at Gohaworki sub-system, the average annual output per unit irrigated land area is 3,300 US$/ha as compared to Godino sub-system with a value of 4,560 US$/ha. The output values per harvested area (sum of areas in a year) is 2,280 US$/ha and 1,650 US$/ha for Godino and Gohaworki sub-systems of Wedecha Scheme respectively.

It can be concluded that land productivity in the community managed schemes is mainly dependent on the following interrelated factors:

- *availability of irrigation water supply*. Whenever the water availability is more dependable/reliable, farmers increase their intensity of irrigation, which increases the annual outputs per irrigated/command area;
- *farm size*. The size of the farm plot is a major factor affecting the willingness of the farmers to invest more on their farm land. It was confirmed that whenever the farm size is extremely small to support an average family size in the area (6 persons household), farmers tend to look for other sources of income. This makes them not to give full attention to their small farming business and not to make optimum investment on it for increased productivity. So, for smallholder farmers, the larger the farm size, the higher the land productivity, taking the extremely small farm size into consideration;
- *willingness to invest more*. Willingness of farmers to invest more on their piece of land depends on the above two factors: water supply dependability and farm size. Whenever farmers are not willing to invest on their piece of land, they also do not want to spend on their plot of land as a full time job.

It is also useful to consider land productivity indicators over consecutive years instead of average values. For this, with a base year of 2007/08 and annual inflation rate of about 10%, both OPUIA and OPUCA were calculated and are shown in Figures 8.5 and 8.6. Crop yield data for the three years were collected from local agricultural development offices and were supplemented with a questionnaire survey. Though the output was a bit higher in 2008/2009 for each scheme, the differences were not so significant over the years. This basically indicates that there is little or no change in the water management aspects to the extent to make significant changes in the land productivity.

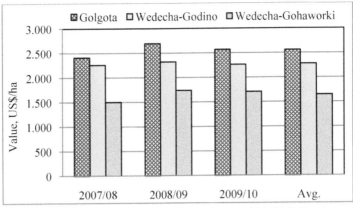

Figure 8.5. Annual output per irrigated (harvested) area (OPUIA) for three consecutive years

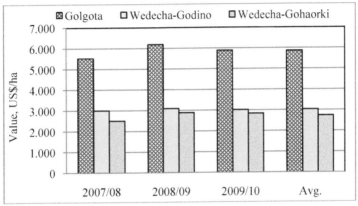

Figure 8.6. Annual output per command area (OPUCA) for three consecutive years

Average annual land productivity at the head, middle and tail reaches of irrigation schemes are indicatives of some aspects of the agricultural system, such as the landholding size and reliability of irrigation water supply. Figure 8.7 shows that the average output per unit irrigated area is particularly higher for the middle reach of Golgota Scheme compared to for the head and tail reaches. This can be well attributed to the larger farm size (average of 1.6 ha at the middle reach) for Golgota Scheme. However, the reach output variations in the other two cases (sub-systems of Wedecha Scheme) cannot be easily attributed to the farm sizes due to more uniform landholding sizes across the reaches.

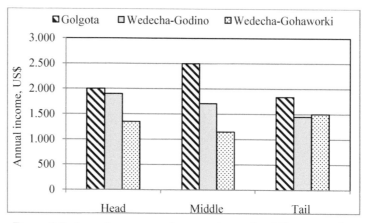

Figure 8.7. Annual output per harvested area for different reaches

Water productivity

The output per unit volume of water becomes important in areas where water availability is a limiting factor or there is anticipated water scarcity. In the two community managed schemes, water is a limited resource, though there are differences in each case. Similarly, access to water is a critical constraint in almost all small-scale irrigation schemes in Ethiopia owing to the poor irrigation infrastructure and water management setups.

Water productivity values were evaluated for each scheme using four different indicators: output per unit irrigation water diverted/supplied, output per unit irrigation water delivered to the field, output per unit water supplied and output per unit water consumed. For agricultural years 2010/2011 and 2011/12, all data of water/irrigation supply were collected. The volumes of irrigation water diverted from the source were measured with stage-discharge (Q-h) relations at a control canal section and Parshall flumes (Figure 8.8). A similar methodology was used for delivered irrigation water at the head of the fields. The consumed water (ET) is the (potential) crop evapotranspiration determined using FAO CROPWAT model Version 8 Swennenhuis (2010b). Irrigation/water supply/delivery components are given in Table 8.4.

Current metering was used at Golgota Scheme to determine discharges at various water depths in the canal diverting water from the river. Control sections were established at two locations: the first just at the downstream of the river diversion and the second at the command inlet. Flow velocities were measured at different verticals in the cross-section of the canal and the 'mean section' method was used to determine the average velocities in different vertical sections. On the other hand, two different sizes of Parshall flumes were used for measuring diverted and delivered irrigation water at Wedecha Scheme. A 15.4 cm (6 inch) and 7.7 cm (3 inch) throat width flumes were used for Godino and Gohaworki sub-systems respectively.

a.*Current- metering for Q-h relation in a diversion canal at Golgota Scheme*
b. *Parshall flume installation at command head at Wedecha Scheme*

Figure 8.8. Measurement of water delivery in irrigation canals

Table 8.4. Annual irrigation/water supply/delivery components in Mm^3

Scheme	Sub system	Annual irrigation water supply	Annual irrigation water delivery	Annual total water supply	Annual water consumed
Golgota		28.71	17.22	30.70	7.43
Wedecha	*Godino*	2.67	1.60	3.32	1.98
	Gohaworki	0.96	0.57	1.15	0.49

Figure 8.9 shows that the output per unit water consumed (OPUWC) is higher than all the other indicators of water productivity except for Godino Sub-system of Wedecha Scheme, which has a higher value for the output per unit irrigation water delivered (OPUID). For Golgota Scheme and Gohaworki Sub-system of Wedecha Scheme, it apparently implies that the volume of water consumed by ET is much less than the diverted/delivered irrigation/water supplies and indicates excess water/irrigation supply.

Comparing only output per unit water consumed (OPUWC) and per unit irrigation water delivered (OPUID) for Golgota Scheme and Gohaworki Sub-system, higher values of the former indicator show that even the irrigation water alone delivered to the command excluding rainfall is much more than the total water demand. Particularly for Golgota Scheme, one can compare OPUWC (0.47 $US\$/m^3$) against OPUID (0.20 $US\$/m^3$) that the consumed water is more than twice more productive than the delivered irrigation water.

The output per unit irrigation water diverted/supplied (OPUIS) and OPUID for Godino Sub-system are more than twice of the corresponding values for Golgota Scheme and Gohaworki sub-system of Wedecha Scheme. It implies that the value of irrigation water is higher for Godino Sub-system implying more productive use of water while irrigation water is least productive at Golgota Scheme. Particularly considering Godino and Gohaworki sub-systems being supplied from Wedecha Reservoir, lower outputs from diverted and delivered irrigation water for Gohaworki Sub-system reveals excess water diversion because of its location of the diversion structure. It also indicates that at Gohaworki Sub-system there is a potential to increase the value of irrigation water by way of saving water.

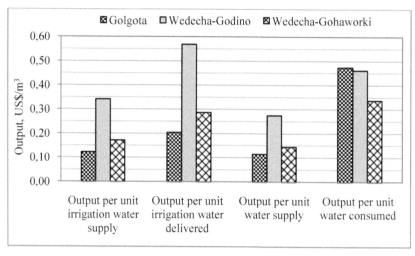

Figure 8.9. Water productivity indicators

While lower values of indicators for OPUIS, OPUID and OPUWS can be attributed to water losses in conveyance, distribution and field application, output per unit water consumed (OPUWC) is not affected by water losses. This is due to the fact that consumed water is the mount which is actually used by ET of the crops. Gohaworki Sub-system has the lowest value of OPUWC, while Golgota Scheme has the highest value. Contributing factors for a higher productivity of consumed water at Golgota Scheme are a set of elements of the farming system, such as soil type, land suitability, crops grown, agricultural inputs and impact of landholding size.

Generally, water productivity at Wedecha Scheme (both sub-systems) is higher than at Golgota Scheme in terms of water (irrigation water) diverted and delivered. Particularly for Godino Sub-system of Wedecha Scheme, water productivity was found to be higher not only in the region but also it is better compared to schemes in other countries such as Hayrabolu, Turkey (Şener et al., 2007); Bhakra irrigation system, India (Sakthivadivel et al., 1999); Mahi Kadana, India, Saldana Colombia, Gorgo, Burkina Faso (Molden *et al.*, 1998). The major issues for higher water productivity at Wedecha Scheme are:

i.   *decision making on irrigation water supply.* The supply (reservoir release) at Wedecha Scheme is based on 'on request' basis. A request for water release is submitted by the WUA to the local agency, which could be modified in the due time. Hence, the water release is scheduled. There is a more demand-based irrigation water supply than at Golgota Scheme (participatory management);

ii. *irrigation water fee.* The irrigation water fee at Wedecha Scheme is area based, in which case there is no way that a farmer pays more for delivering more water than he actually needs. Still, it was observed that whatever may be the assessment method, irrigation water fee is a real incentive for saving water.

It is useful to observe that the dependability of irrigation water supply while it can potentially increase land productivity, there should be a mechanism for controlling excess beyond a certain level of dependability. In the absence of the supply monitoring, it can significantly decrease water productivity.

Godino Sub-system water productivity indicators for OPUIS, OPUID and OPUWS can be best benchmarked to other schemes where water is scarce or access to water is constrained. On the other hand, Golgota Scheme indicator for OPUWC can be used as a benchmark for water productivity improvement activities in the region through better soil and crop management practices in addition to water management.

8.2.3 Physical indicators

Lack of sustainability has been a critical issue in the community managed small-scale irrigation schemes in Ethiopia since long. Issues constraining sustainability in these schemes are in general a collective result of non-sustainable irrigation water supply, defective irrigation scheduling, size of landholding, decreasing land productivity, etc. In fact, sustainability in small-scale irrigation schemes in Ethiopia is a key issue for food security and rural livelihood enhancement as it comprises about 70% of the total irrigation development.

Data on three different sizes of land related to the schemes were collected to evaluate the physical indicators, i.e. irrigable land, initially irrigated land and currently irrigated land. The irrigable land of each scheme/sub-system was determined by surveying the boundary of the command areas using Global Positioning System (GPS). These were then added to GIS where the boundaries were plotted and the areas were determined. The initial irrigated areas when each scheme was commissioned were taken from the scheme design reports and the same were confirmed from local irrigation agencies with some adjustments. The adjustments were required because the whole designed area might not have been fully irrigated when the scheme was commissioned.

'Currently irrigated areas' for each scheme were determined in two ways. First, at each scheme there is an up to date record of the members of the WUA along with their irrigated landholdings, complied by the schemes' water users associations. So, the irrigated area at each scheme was found out as a sum of the irrigated holdings of all farmers belonging to the water users association. Second, with the boundaries of total irrigable command of each scheme determined, a survey was also conducted using GPS to determine non-irrigated lands, residential areas and grazing land areas. The net irrigated land area was then determined as the difference between total command area and sum of all non-irrigated land areas within the command. The irrigated area at each scheme remained the same over the years 2007 through 2010. Land areas pertaining to the schemes and physical indicators are given in Table 8.7.

Table 8.7. Physical performance indicators (2007-2010)

Scheme	Sub-system	Irrigable land, ha	Initial irrigated land, ha	Current irrigated land, ha	Indicators	
					Irrigation ratio	Sustainability of irrigated area
Golgota		600	450	550	0.92	1.22
Wedecha	*Godino*	300	250	200	0.67	0.80
	Gohaworki	60	60	50	0.83	0.83

Irrigation ratio, being an indicator for the degree of utilization of the available land for irrigation, is also a useful indicator for whether there are factors contributing for under irrigation of the command area. Irrigation ratio is higher for Golgota Scheme with a value 0.92 (92% of the irrigable command area is currently under irrigation) and Godino sub-system with the lowest value. Greater irrigation ratio at Golgota Scheme can be explained by three factors: namely generous water availability, absence of irrigation water fee and better land productivity.

Lower irrigation ratio at Godino Sub-system of Wedecha Scheme is attributed to lower reliability of irrigation flows during some months of the year, irrigation water fee charged by the regional irrigation authority and relatively lower land productivity compared to Golgota Scheme. The combination of these three factors at Godino Sub-system has increased the amount of non-irrigated land and hence the irrigation ratio. The non-reliability of irrigation supply to Godino sub-System is aggravated by two factors: first due to the 'on request' based water supply system and second its location of diversion being located on the downstream of Gohaworki sub-system (Dejen et al., 2012).

Irrigation ratios in these two schemes are higher compared to similar schemes in other parts of Ethiopia. Awulachew et al. (2005) reported that small-scale irrigation schemes in Ethiopia on average perform at about 40% on average in terms of irrigation ratio. In the schemes under this study, relatively good access to market and practice of marketable crops enabled a generally higher irrigation ratio. Şener et al. (2007) also presented irrigation ratios for Hayrabolu irrigation scheme in Turkey over 16 years where the average value was 27%, in which case the schemes under the current study perform much better. But, the comparison between the two schemes is more important than comparison with national average performances or international cases.

Sustainability of irrigated area, which tells on whether the area under irrigation is contracting or expanding right from the commencement of the scheme, is a useful indicator for sustainability of irrigation. In Ethiopia, community managed schemes suffer severe non-sustainability problems ranging from complete abandonment of schemes to different levels (Awulachew et al., 2005).

Sustainability of irrigated area in Godino and Gohaworki sub-systems of Wedecha Scheme have more or less similar values, 0.80 and 0.83 respectively implying reduction of irrigated areas by about 20%. Still, these contractions in the irrigated areas are minimum compared to schemes in the region. For Golgota Scheme with a value of 1.22, the irrigated area has expanded by about 20% since commissioning. Awulachew et al. (2005) stated that the major issues responsible for non-sustainability (contraction of irrigated areas) in small-scale irrigation schemes in Ethiopia are conflicts in water use, lack of sense of ownership, non-sustainability of water sources, poor institutional setups for water management, etc. These authors indicate that expansion of irrigated area at existing small-scale schemes is seldom observed in some successful schemes.

Expansion of the irrigated area at Golgota Scheme has to do with some issues pertaining to the scheme as follows:

i.   *better reliability of the irrigation supply*. The irrigation supply as stated is highly reliable at this scheme, though that results in a lower water productivity. This encourages farmers to irrigate their piece of land fully;

ii.   *absence of irrigation water fee*. Absence of irrigation water fee attracts more farmers to this scheme to look for pieces of uncultivated land. Moreover, the more generous water supply for free is an incentive for large numbers irrigators to move to the scheme for leasing land in the scheme;

iii. *capacity of the WUA*. The overall capacity and flexibility of the WUA at Golgota Scheme is high. This includes responsiveness of the WUA to water requests, freedom of access to water in a more flexible way and better capacity for maintenance.

8.2.4   Monthly comparison of water supply indicators

While the annual water supply indicators are useful for aggregated water supply/demand of the scheme, they do not indicate the specific periods in a year with excess/shortage of water/irrigation supply. So for each scheme, monthly values of irrigation/water supply/demand were determined for monthly indicators. Monthly water demands are potential ET values. They were determined based on climate data, cropping pattern and crop data using FAO CROPWAT 8.0 (Swennenhuis, 2010a) for each scheme. Monthly irrigation demands for the schemes were also determined using FAO CROPWAT 8.0 as a difference between monthly water demand and effective rainfall. Monthly irrigation supplies were determined by continuous flow measurement of irrigation water delivery at the inlets of the command areas over a period of two years using Parshall flumes or Q-h relations. Monthly water supplies were then determined as the sum of monthly irrigation supplies and effective rainfall. The monthly water/irrigation supply/demand components and indicators are given in Figures 8.10 through 8.15.

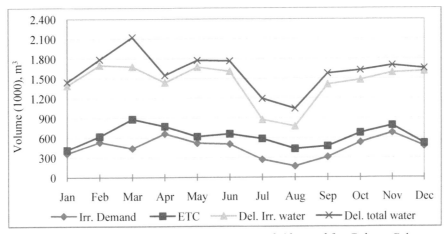

Figure 8.10. Monthly total water/irrigation supply/demand for Golgota Scheme

It is observed from Figure 8.11 that for Golgota Scheme the monthly RIS are higher than RWS for all months; which implies that irrigation is the major source of excess supply at the Scheme. Moreover, it is evident from the figure that the RIS values are variable throughout the year. Monthly fluctuation of RIS implies the weaknesses of irrigation scheduling and supply. At Golgota Scheme it confirms that irrigation water diversion at Golgota is based on stage of water in the river and does not well address demands. All valves of RIS being higher than 2.0, though it indicates excess water supply throughout the year, exceptionally high irrigation supplies occur during the months of August, September, January and March. There is significant amount of rainfall during July, August September, with high river stages. While there is practically very little irrigation demand during these months, farmers still divert water and it is released at the tail end of the command. Moreover, farmers keep on diverting water during off irrigation periods (based on cropping patterns), where the field demand significantly falls, which intermittently leads to high RIS.

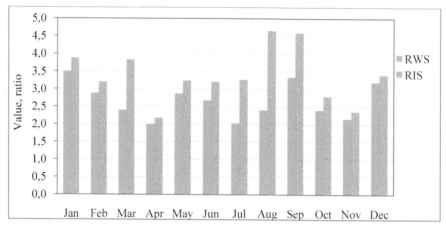

Figure 8.11. Monthly water supply indicators for Golgota Scheme

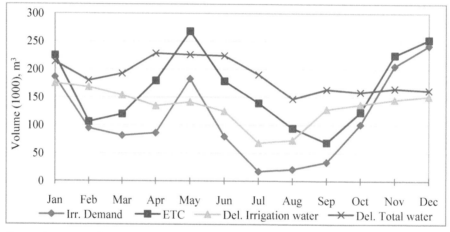

Figure 8.12. Monthly total water/irrigation supply/demand for Wedecha (Godino sub-system)

In addition to the annual water and irrigation supply, monthly water supply indicators help identification of the specific months in a year that need attention with regard to water management. Modification in a few months of the year can result in considerable saving on irrigation water. For the case of Golgota Scheme, these are either main rainy seasons or months of low irrigation demand (January-February. Irrigation water diversion in a majority of community-manages schemes in Ethiopia is supply-based, which means the management lacks ensuring supply-based diversion. Hence, demand-based water diversion during off (minimum) irrigation is crucial for both saving water and environmental sustainability in these schemes.

For Wedecha Scheme (Godino sub-system) (Figure 8.13), monthly RWS and RIS are closer to each other with the exception of the months July, August and September, where RIS values are too high. At this scheme more than 60% of the annual rainfall occurs during these three months and there is practically very low irrigation demand. Although the irrigation agency controls the water diversion, it was not effective enough to keep the RIS uniform throughout the year. High RIS during the rainy months actually show the poor management capacity of the agency, because the major management objective is to supply the right amount of water at the right time. The RIS also tends to be higher during February to April which is the minor rainy season of the area. It is also worth to see that unlike the fact that annual RIS and RWS values are higher than 1.0, there occurs water stress during the dry months of November to January and May where both monthly indicators are lower than 1.0. These months are actually months of high irrigation requirements and need emphasis to smoothen the irrigation supply throughout the year. Hence, while annual RIS and RWS can give a general view of the water supply, monthly indicators are more useful.

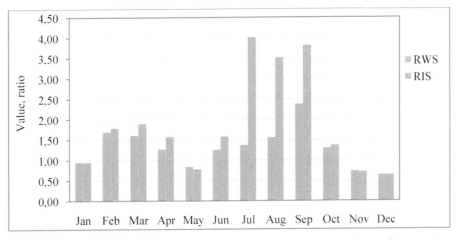

Figure 8.13. Monthly water supply indicators for Wedecha (Godino sub-system)

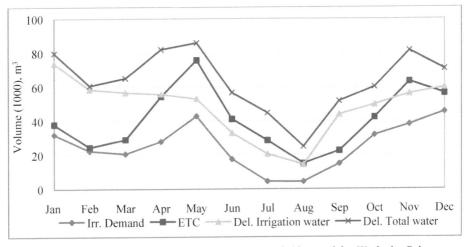

Figure 8.14. Monthly total water/irrigation supply/demand for Wedecha Scheme
(Gohaworki Sub-system)

The monthly values of indicators for Gohaworki sub-system (Figure 8.15) are variable throughout the year. However, unlike Godino Sub-system, for Gohaworki Sub-system both indicators are higher than 1.0 for each month depicting demands are met throughout the year. The RIS is much higher during the main rainy months of July to September, the same as Godino sub-system. Relatively higher values of indicators were also observed during January to March. The fact that Gohaworki sub-system diversion structure is located on the upstream of Godino Sub-system enabled it to deliver supplies sufficient to meet demands throughout the year. However, at both sub-systems, matching supplies with field demands is the main concern to flatten the monthly fluctuation of water supply indicators thereby saving irrigation water lost during off-irrigation months and due to excess supply during irrigation months.

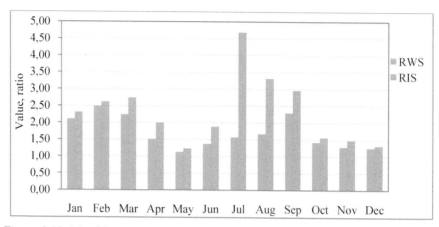

Figure 8.15. Monthly water supply indicators for Wedecha Scheme (Gohaworki Sub-system)

8.3   Major issues for adequate irrigation management in the community managed schemes

Comparative performance assessment at Golgota and Wedecha community managed schemes enabled identification of some crucial water management issues to address important aspects of water saving, productivity and sustainability in these schemes. The following five factors have been identified as the most important ones.

8.3.1 Decision making on water diversion

There have been several positive experiences from participatory irrigation management around the world during the past two decades. Irrigation management by the water users themselves is particularly important in Ethiopia for two reasons: First, public institutions for small-scale irrigation management are generally weak and incapable. Second, absence of sound irrigation infrastructure for easy flow control and monitoring in these schemes makes it difficult for irrigation management institutions to make adequate management. Third, here is a weak linkage between water users and institutions for irrigation management.

In the specific cases of the schemes of this study, irrigation water management solely by the water users at Golgota Scheme was the reason for a higher land productivity and expansion of the irrigated area in the scheme. However, it was at an expense of large water losses and low water productivity. Moreover, waterlogging is another threat in this scheme being without any incentive to save water. On the other hand, decision making on water diversion by an agency at Wedecha Scheme was the reason for lower land productivity and shrinking irrigated land, though it has a higher water productivity.

8.3.2   Irrigation water fee

It is only very recently in Ethiopia that farmers pay for irrigation water. Agricultural water had been a free commodity until recently particularly in schemes for food production. This had positive results for the country's food security through expansion of small-scale irrigation for free access to water for smallholder farmers. However, free access to water cannot be the best solution for sustainable irrigation development. In view diminishing water resources and need to improve water productivity, appropriate irrigation water fee is a good incentive for saving water and ensuring environmental sustainability.

8.3.3   Capacity of the WUA

While water user associations have proven to be good options for managing irrigation water in several community managed schemes in Ethiopia, they lack the capacity to adequately implement the management. Weaknesses of the WUA in these schemes include lack of adequate technical and institutional capacity. Although farmers had long experiences with traditional irrigation, they have little experiences in handling modern irrigation infrastructure. With the introduction of modern irrigation facilities and irrigation management transfer to WUA, there lacks a training to water users regarding infrastructure management, water management, water conflict resolution, etc. To this end, training focussed on overall capacity building of the WUA is a prerequisite for sustainable irrigation management.

8.3.4   Condition of land and water resources

The condition of the water source plays a vital role in land and water productivity and sustainability of community managed irrigation schemes. Whenever water is more constraining than land for production, the competition for water increases, in which case water productivity and saving becomes more important. For the actual condition of the schemes in this study and other similar schemes, the WUA lack the overall knowhow to save the limited water resource. This hence results in not only a low water productivity, but also low land productivity. In this case, an involvement of an external agency is a good option to achieve the objective of higher water productivity through more efficient water management. In this point of view, in the case of Wedecha Scheme water is more limiting than land and participatory water management (involvement of an agency) is appropriate. However, the involvement would not have to be to the extent to make the irrigation supply unreliable. On the other hand, when land is a more limiting factor than water as in the case of Golgota Scheme, irrigation water management exclusively by the water users is appropriate; because in this case water mismanagement cannot be a cause for lower land productivity. In this, land productivity is rather related to other agricultural practices.

The key operational factors along with positive and negative implications on the external (comparative) performance in the community managed irrigation schemes are summarized in Table 8.5. The proposed step in decision making of an appropriate irrigation management setups in community managed irrigation schemes is shown in Figure 8.16.

Table 8.5. Key operational issues for land and water productivity and sustainability

Factor	Conditions	Key issues	
		Positive	Negative
Decision making on water delivery	External agency	• Higher water productivity	• Lower RWS and RIS • Lower land productivity • Contraction of irrigated area
	Water users	• Higher confidence on water supply • More sustainable irrigated area (expansion)	• Lower water productivity • Threat of waterlogging (less sustainable)
Irrigation water fee	Free access to water	• Less conflict on water	• No incentive for water saving:- waste • Lower water productivity • More conflict on water
	Fee paid for water management	• Water saving and hence higher water productivity • Quicker response on major maintenance requests	
Autonomy of WUA	Autonomous	• Expansion of irrigated area	• Lower water productivity
	Non-autonomous	• Higher water productivity	• Less sustainable irrigated area • More conflicts on water

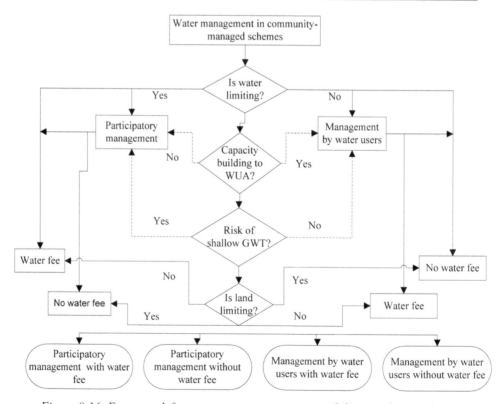

Figure 8.16. Framework for water management responsibilities and water fee

9 INTERNAL EVALUATION OF IRRIGATION SERVICE IN THE COMMUNITY MANAGED SCHEMES

9.1 Rationale for internal irrigation performance evaluation in the community managed schemes

Internal irrigation performance assessment is related to internal management processes and irrigation services. Irrigation schemes in general vary in their internal processes and features such as in institutional setups, infrastructure, source of water, method of water acquisition, means of water distribution, field irrigation, etc. Golgota and Wedecha community managed irrigation schemes also vary in several internal processes. Owing to basic differences in the source and condition of water and institutional setups, the internal management targets of each of these schemes is different. Internal (process) performance assessment in these schemes enable evaluation of the levels of irrigation services whose targets are scheme specific and depend on water users expectations.

In the community managed schemes of this study, while comparative performance evaluation is of high significance for cross comparison for scheme water supply, land and water productivity and irrigated area sustainability, it is less useful for directly addressing the internal variations in performance. Internal irrigation performance in these schemes is also a concern, because internal processes and practices related to supply scheduling, time of water delivery and flow rates within each scheme (at different levels and user groups) are important factors from the point of view of service provision. Internal process indicators can provide a more detailed acquaintance with the nature of the management processes and practices responsible for a given irrigation service level. Moreover, they enable identification of the things to do for improvement of internal and overall performance. The purpose of internal (process) performance assessment in these schemes is to appraise the irrigation service level from water users' views in the absence of qualitative data on water deliveries.

9.2 Irrigation service: water users' views

It is well articulated that although there are several stakeholders involved in irrigated agriculture, farmers are often given the least attention in all phases from planning to operation and performance monitoring of irrigation schemes. Farmers are the end recipients of the irrigation service and are the ones consuming the water. It needs to be stressed that the ultimate purpose of the whole business is to boost agricultural produce and farmers are responsible for it. It is now well understood that the success and sustainability of community irrigation schemes by in large depend on adequate involvement of water users in planning, development, operation and performance evaluation. Framers can have different views into the irrigation services they get, and hence interventions by other stakeholders may not necessarily be appropriate for them. Irrigation service is generally based on service delivery arrangement (Burt and Styles, 1998). After an irrigation service agreement specifying the water rights, water fee arrangements, location of the water delivery, degree of flexibility of the supplies, etc., performance monitoring follows. This performance monitoring would need in particular to integrate the water users' perceptions.

Internal irrigation service evaluation needs to be considered as an integral part of the operation and management of irrigation schemes. Evaluation of internal irrigation performance indicators generally requires quantitative data on water deliveries, which is not a priority in community managed schemes in least developed countries in general. In these smallholder community managed irrigation schemes, water delivery service plays a vital role to enhance their productivity. While irrigation performance evaluations in the past have generally addressed the needs of several stakeholders, in these evaluations, farmers' views have not been taken well into account. There is only limited literature on irrigation performance evaluation from the farmers' perceptions in the absence of other data. Svendsen and Small (1990) and Sam-Amoah and Gowing (2001) suggest that farmers are the final consumers of the services and irrigation service evaluation would have to address their need.

In the Golgota and Wedecha community managed schemes, no quantitative flow data are available on water deliveries at various delivery points. Although there were some irrigation performance evaluations so far in these schemes, they only focused on the external indicators, and farmers perceptions and variations in the service levels across different groups of users and their causes were overlooked. In these schemes, issues related to water rights, water sharing, time of delivery, flow rates and flexibility of water delivery schedules are the major concerns. However, for the farmers of these schemes, the flexibility of the supply is more important than the issues of water rights. This is because farmers don't care much for the total volume of water they receive per cropping season; but they do care more for the appropriateness of the delivery based on their needs. In the absence of data on irrigation flows, the quality of irrigation services called utility have been determined from qualitative farmers' responses.

9.3   Factors considered for utility of the water delivery service

Utility is composed of elements of the service delivery system. The condition of these elements of the irrigation service to farmers enables evaluation of the utility of the service. The factors selected for utility can be different from on scheme to another based on the specific situation of the irrigation service delivery system. Three important factors: timing, dependability and tractability were considered to measure the utility in this case. Two sub-factors were considered under each utility factor to make it easier for the farmers to explicitly respond to the questions related to these factors (Table 9.1).

- *Timing*. Refers to the suitability of the time of water delivery to the users. Farmers have their own preferences of time for field irrigation. This refers to the timing during a day or flexibility in the date of irrigation. It also includes the duration of time for which water is supplied;
- *Dependability*. It is for the farmers' level of confidence on availability of the water supplies as planned. It also refers to the knowledge of the future water supply schedule and its uncertainty. Predictability is particularly very important for sustainability of the community managed schemes of this study. It also has implications on the use of inputs and farmers' investments on their piece of land;
- *Tractability*. It implies the capacity of the farmers to effectively make control of the water supply and irrigate their piece of land. The size of the steam size of the water supply in community managed schemes is often variable. In the community managed schemes of this study tractability is also a fundamental element of the water supply.

Table 9.1. Factors and sub-factors considered for utility

Utility factor	Sub-factors
Timing	water arrival time
	duration of delivery
Predictability	knowledge of delivery
	confidence of availability
Tractability	stream size
	point of delivery

9.4 Fuzzy set and fuzzy linguistic terms considered

Often, the classes of objects encountered in the real world including the current study on irrigation service levels do not have precisely defined criteria of membership. Moreover, human perceptions (attitudes) often have ambiguous boundaries and are fuzzy. A set of fuzzy classes cannot be described in the classical sets. Yet, the fact remains that such imprecisely defined 'classes' play an important role in human thinking, particularly in the domains of pattern recognition, communication of information, and abstraction (Zadeh, 1965).

A Fuzzy Set is a set in which the elements of a set have partial membership in the set. Unlike classical sets, fuzzy set theory permits the gradual assessment of the membership of elements in a set, which can be described with the aid of a membership function $\mu(x)$ in the real unit interval $[0, 1]$. Fuzzy sets are fully defined by the membership function and are useful for handling vagueness and imprecision. If U is a collection of objects denoted generically by x, a fuzzy set A is defined as:

$$A = \left\{ \left(x, \mu_A(x) \right) | x \in U \right\} \tag{9.1}$$

The flexibility in fuzzy sets allows mathematical representation of non-precise human concepts and enables applications in a variety of fields. In this study, fuzzy sets are suitable for representing the opinions of the farmers' on the irrigation services they receive. Fuzzy linguistic variables were used to collect data on the irrigation services. A fuzzy linguistic variable is a variable whose domain is a collection of pre-specified fuzzy concepts (Ngan, 2011). These variables are generally used in a variety of surveys or questionnaires. In this study, seven fuzzy variables were considered for appropriateness of irrigation service criteria; namely very good, good, more or less good, medium, more or less bad, bad and very bad. Similarly, seven variables for importance (weight) of service criteria were considered: very high, high, more or less high, medium, more or less low, low and very low. The support functions employed for the seven variables are as given in Figure 9.1.

9.5 Data for utility and fuzziness of responses of farmers

Evaluation of the quality of the irrigation services in these schemes is based on qualitative data of farmers' perceptions. Farmers provided their linguistic evaluations on the levels of the irrigation services they receive. Data on the service levels with regard to the utility factors were collected from the water users using a questionnaire. It is likely that an irrigation service favours certain groups of water users than others. Thus, the degree of satisfaction and hence their evaluation would be variable across different groups of water users (head, middle and tail). Hence, a sample of 10 farmers from each reach based on their location and a total of 30 farmers for each scheme/sub-system of a scheme were selected for an interview. Data on the suitability of each utility factor and its importance to them were collected in the form of linguistic expressions. These expressions are associated with fuzziness and clear distinctions are difficult to make easily.

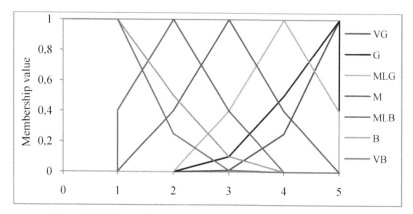

Figure 9.1. Seven linguistic terms used and support functions

Hence, the linguistic expressions of farmers' perceptions were represented by fuzzy sets with 5 elements. The farmers' fuzzy expressions have supports as shown in Table 9.2 for the appropriateness and importance of the utility factors to the farmers. The expressions of all water users regarding each sub-factor of utility within a stratum of water users were aggregated to a single linguistic expression, which can be represented by a single fuzzy set. The resulting fuzzy set can be normalized by dividing all the elements by the maximum support so that at least one element of the set has full membership. Generally, a fuzzy set resulting from the application of set operations cannot be easily approximated to the linguistic expressions. Hence, the set can be made convex by changing some of the elements.

Table 9.2. Fuzzy linguistic expressions with supports (Svendsen and Small, 1990)

Expressions	$\mu(1)$	$\mu(2)$	$\mu(3)$	$\mu(4)$	$\mu(5)$
Very good/very high	0	0	0.01	0.25	1
Good/high	0	0	0.1	0.5	1
More or less good/more or less high	0	0	0.4	1	0.4
Medium	0	0.4	1	0.4	0
More or less bad/more or less low	0.4	1	0.4	0	0
Bad/low	1	0.5	0.1	0	0
Very bad/very low	1	0.25	0.01	0	0

However, it is more convenient for system managers and other stakeholders to use this measure of the level of service when it is converted into a numerical value. This numerical indicator is called farmer utility (FU). Utilizing the same method as was described by El-Awad (1991) for a universe of five elements and support functions, the FU can be evaluated from:

$$FU = \{1/(N-1)\} * [\{\sum(i-1) * \mu_i\}/\sum \mu_i]$$

(9.2)

Where i is possible values of the universe U and μ_i is the support (membership) of the element i of the fuzzy set. This numerical utility indicator ranges from 0 to 1, with higher values indicating better irrigation service level.

9.6   Results of utility analysis

9.6.1   Utility for importance (weight) of factors

Importance refers to the relative significance to the farmers of the utility factors ranging from very low to very high in the form of linguistic expressions. Farmers' responses on the significance of each factor were aggregated to a single linguistic expression and utility indicator. First, the utilities for the importance of the sub-factors were evaluated. The arithmetic mean of the weight (importance) of the sub-factors yielded the utility (weight) of the main criterion (timing, dependability and tractability) for each scheme. Higher utility values refer to more significance attached to the utility factors. The utility for the significance of the factors was determined for each reach as shown in Table 9.3.

Table 9.3. Utility for importance of water delivery factors to farmers

Utility factor	Golgota		Wedecha-Godino		Wedecha-Gohaworki	
	Utility	Expression	Utility	Expression	Utility	Expression
Timing	0.68	High	0.55	Medium	0.62	More or less high
Dependability	0.52	Medium	0.85	High	0.73	High
Tractability	0.72	High	0.76	High	0.71	High

Figure 9.2. Flow at the head of an irrigation block at Golgota Scheme (lacks adequate farm structures for delivery to farthest outlets)

Table 9.3 depicts that the utility factors in the order of their importance at Golgota Scheme are dependability, timing and tractability, with dependability being the least important factor. There is a rotational delivery schedule via three offtakes on which all the water users at this scheme have agreed. This interestingly showed that the water delivery at this scheme is predictable enough for the farmers to attach a lower utility of importance to predictability. Farmers had concerns on the timing of delivery as they have to irrigate the whole day on their turn. Moreover, they are required to irrigate on holidays and Sundays as long as that is their turn. The highest utility of importance was attached to tractability, particularly to the point of water delivery. Water is delivered through a single offtake at the head for over 100 water users (Figure 9.2). Within the irrigation unit, it poses problems in water distribution and sharing due to lack of adequate farm structures and distances to some farm outlets.

For Godino sub-system of Wedecha Scheme predictability is the most important followed by tractability and timing. Aggregated expression of farmers' attitudes for both predictability and tractability is 'high'. There is an established water delivery schedule; however, water users attached a higher utility to these factors. Farmers were very much concerned to the fact that the decision on water diversion is not in their hands, in which case they attached a higher utility of importance to predictability. Moreover, they also expressed that if water has been delivered, its tractability is important to them as well. Farmers were however not worried much about the time at which water is delivered as long as it is during day hours.

For Gohaworki sub-system, farmers attached the same orders of importance for the factors as for Godino sub-system. However, the utility values for the importance of the factors for this sub-system are lower than that of Godino sub-system particularly for the two important factors (Table 9.3). Less importance of predictability for Gohaworki than Godino sub-system can be attributed to its location of water diversion being on the upstream, which is not very surprising. Most important is that the utility for the importance of factors depends on the service setting and management responsibilities at different levels (tractability for Golgota Scheme and dependability for Wedecha Scheme).

9.6.2 Utility for service rating (appropriateness)

The utility for the services was assessed for each scheme within the head, middle and tail reaches with respect to each utility factor/sub-factor. First utility with respect to the sub-factors was evaluated for appropriateness and then aggregated to the utility of the three factors. Table 9.4 shows the utility for Golgota Scheme within each reach.

Table 9.4. Farmers' utility for Golgota Scheme

Reach	Timing		Dependability		Tractability	
	Time of water arrival	Duration of delivery	Knowledge of future delivery	Confidence on delivery	Stream size	Point of delivery
Head	0.72	0.64	0.73	0.71	0.60	0.62
Middle	0.83	0.75	0.72	0.66	0.73	0.72
Tail	0.68	0.64	0.71	0.56	0.67	0.70

For the most important utility factors of tractability and timing, farmers in the middle reach recorded the highest utility values for Golgota Scheme (Figure 9.3). The lowest utility for the head reach was of tractability. Head reach farmers had difficulty in delivering water from the main canal due to absence of appropriate farm turnouts and effect of sedimentation, and hence had the lowest tractability. On the other hand, the lowest utility for the tail reach was of dependability. Although farmers knew the schedule for their turn of delivery, they recorded a lower utility for a confidence on availability as per the schedule than middle and head farmers.

The utility for Wedecha Scheme (combined sub-systems) with respect to each sub-factor is given in Table 9.5. Head users at Wedecha Scheme recorded the highest utility for the appropriateness of the services with respect to all factors followed by middle and tail users. While this is not unexpected in such schemes, the farmers in the tail reach had a low assessment for all the factors unlike in the head and middle reaches (Figure 9.4). It is notable that for the most important factor of dependability at this scheme, water users in all the reaches had a utility of 0.5 or lower, confirming that this is the major concern of the farmers. For the less important factors of tractability and timing at this scheme, however, head and middle users recorded utility values higher than average.

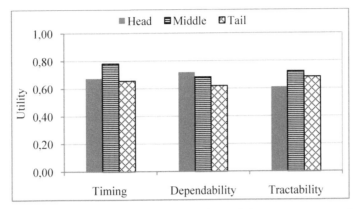

Figure 9.3. Utility at Golgota Scheme in each reach for service appropriateness

Table 9.5. Farmers' utility for Wedecha Scheme

Reach	Timing		Dependability		Tractability	
	Timing of water arrival	Duration of delivery	Knowledge of delivery	Confidence on delivery	Stream size	Point of delivery
Head	0.73	0.72	0.5	0.50	0.71	0.71
Middle	0.69	0.69	0.5	0.40	0.56	0.67
Tail	0.45	0.40	0.5	0.34	0.36	0.54

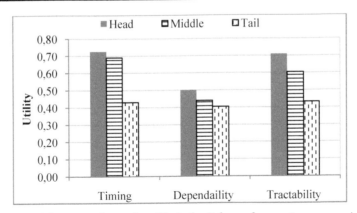

Figure 9.4. Utility in each reach at Wedecha Scheme for service appropriateness

9.6.3 Overall utility

The overall utility of irrigation service helps to quickly see the mean overall assessment of the water users. This in principle has to involve computation of fuzzy weighted mean of farmers' opinions on the service. A simplified method of assessment was employed in this case, in which the aggregated utility for service appropriateness and importance is considered:

$$M = \frac{\sum_1^N A_i * W_i}{\sum_1^N W_i}$$
(9.3)

Where M is the weighted mean of farmers' evaluations of the service, A_i is an aggregated utility for a particular factor of utility (criteria), W_i is the weight attached to the utility factors, and N is the number of utility factors considered in this case. The aggregated utility of each factor was as determined based on sub-factors under each main factor. The utility for the importance of the factors to the farmers (Table 9.3) is nothing more than the weight of each utility factor. This enabled calculation of the overall utility within each reach from the three main factors.

Table 9.6 shows the weighted overall utility in each reach for Golgota Scheme, in which the weight was based on a qualitative relative importance attached by the farmers to the sub-factors. The overall utility in the head and tail reaches was lower and same, while it was higher in the middle reach. The irrigation service was 'more or less good' in the head and tail reaches; while it was 'good' in the middle reach. Hence, even if water users in different reaches attach different importance (weight) to different factors and have different perceptions on the appropriateness of each factor, the overall irrigation service utility can be similar. But the utility of the individual factors assist decision making for a better service. At Golgota Scheme, water management measures need to address particularly tractability and dependability in the head and tail reaches respectively.

The overall weighted utility for Wedecha Scheme is shown in Table 9.7. Head users had the highest overall utility followed by middle and tail users. Particularly the overall irrigation service level in the tail reach was more or less bad with an overall utility of 0.42 and practically all the factors had more or less similarly contributed to it. On the other hand, in the head and middle reaches, dependability was a greater contributor for a lower overall utility. Water management interventions for Wedecha Scheme would have to address dependability in all the reaches. Moreover, all the factors would need to be addressed in the tail reach.

Table 9.6. Weighted average overall utility at Golgota Scheme

Factor	Timing	Dependability	Tractability	Overall utility
Weight (utility for importance)	0.68	0.52	0.72	
Head	0.67	0.72	0.61	0.66 (more or less good)
Middle	0.78	0.68	0.73	0.73 (good)
Tail	0.66	0.62	0.69	0.66 (more or less good)

Table 9.7. Weighted average overall utility at Wedecha Scheme

	Timing	Dependability	Tractability	Overall utility
Weight (utility for importance)	0.59	0.79	0.74	
Head	0.73	0.50	0.71	0.66 (More or less good)
Middle	0.69	0.44	0.60	0.59 (average)
Tail	0.43	0.40	0.43	0.42 (More or less bad)

The objective of these community managed irrigation schemes is to enhance the country's food security for households with small landholdings. The success of these schemes is strongly dependent on the quality of irrigation services. Some of the internal water management practices responsible for ill-utility at Wedecha Scheme include:

- *institutional arrangement for water management*. The trend in irrigation service levels at Wedecha Scheme from the head to the tail users could be expected in several of such irrigation schemes. However, irrigation management would need to aim to implant equity to users. Institutional setups for decision making on water release, water sharing and delivery determines the utility. The irrigation water supply arrangement of Wedecha Scheme is 'On request' type; where water users submit their water need through their WUA and the local agency makes the water release. Irrigation needs are being submitted together with other irrigation districts' water needs, which the agency usually fails to follow the schedule for the water needs. It made farmers of the scheme to attach a higher importance to dependability of the supply and a lower utility to it. Supplies frequently go off-schedule and less predictable. Farmers prefer transfer of the whole irrigation management to themselves; however, it is unlikely that they manage it in a sustainable way given the poor capacity and institutional strength of the WUA. To, this end, the local public agency needs to improve its responsiveness through service-based staff training and capacity building. Moreover, the relationship between the WUA and the agency needs to be strengthened;

- *failure and mis-operation of water division and farm structures*. Lack of adequate farm and off-farm infrastructure for water control and distribution across different groups of users is a foremost aspect that resulted in a low utility particularly in the tail reaches. This concerns structures such as diversion offtakes, division boxes and farm turnouts. These structures were not only insufficient but also mis-operated and the gates damaged (Figure 9.5). This apparently gave the head users a comparatively more reliable access to water. In terms of the timing and tractability of the supply, it also favoured the head users, because they denied users down the system from accessing water. Hence, tail users often got water at inappropriate timing and at inappropriate flow rates. This was also evident from the lower utility in the tail with respect to all factors. Asset management is a core in sustainable irrigation management; however improving the physical infrastructure alone cannot achieve sustainability. Thus, infrastructure and the required expertise of the users for its proper use are inseparable;

Figure 9.5. Water division structures with demolished gates at Wedecha Scheme

- *implementation incapability of the WUA*. The overall capacity of the WUA in implementing decisions and activities related to irrigation water management is critical for irrigation service utility. These include appropriate decision making and implementation on: illegal acts such as water denial on the downstream, unscheduled water delivery, damage of structures; planning annual and major maintenance; planning routine maintenance; monitoring service for continuous improvement; etc. At Wedecha scheme, the WUA had an extremely limited capacity for keeping the overall service delivery healthy and apparently monitoring and feedbacks were absent. The incapability of the WUA was fundamentally technical, followed by financial. The association had long experience with traditional irrigation that didn't require sound and collective decision making on water management. However, training on capacity building on service oriented management of irrigation was absent and had limited finance for running their day-to-day activities. The WUA needs to be capacitated to effectively carryout regular monitoring of the services and implement regulatory measures. The local public irrigation agency would have to take the initiative for the capacity building. Currently, the farmers pay a water fee to the agency, and not to the WUA. Financial capacity can be strengthened through charging a reasonable annual fee for routine maintenance and day-to-day water management to be collected by the WUA in addition to the annual water fee for the irrigation agency.

9.7   Utility and productivity

The ultimate purpose of good irrigation service provision and hence high utility is to increase agricultural productivity. Hence, it will be interesting to evaluate the relationship between utility and productivity (yield). Of course, land productivity is a function of several factors of the agricultural system in addition to water supply including use of agricultural inputs, soil types, land holding size, etc. Water availability, however, remains to be an essential element without which cropping could have been possible only during a rainy season of 4 months. Owing to the better and more uniform utility levels across the reaches at Golgota Scheme, there appeared no relationship between the utility and productivity (Table 9.8). Farmers were if not fully, in majority satisfied with the irrigation service at Golgota Scheme, and the effects of the other factors for productivity were not visible. On the other hand, at Wedecha Scheme, visible differences in the overall utility across the reaches resulted in a corresponding difference in yield. This depicts that the effect of utility would be visible on farmers' yields when the differences in utility are considerable and that the effect of other factors than water on yield are far lower that the effect of utility.

Table 9.8. Relationship between overall utility and agricultural output

Reach	Golgota		Wedecha	
	Overall utility	Average productivity (US$/ha)	Overall utility	Average productivity (US$/ha)
Head	0.66	2,670	0.66	2,400
Middle	0.73	2,670	0.59	2,060
Tail	0.66	2,670	0.42	1,700

9.8   Conclusion

In the majority of the irrigation schemes in least-developed countries, data of actual water deliveries are not readily available, and is not a priority either. Particularly in many of the community managed schemes, installation of flow measuring structures is not considered due to the nature of the schemes; and those existed in some of these schemes were not properly used. Thus, the conventional internal water delivery performance indicators that depend on water deliveries to different delivery points cannot be used. The qualitative approach for irrigation service evaluation that was used here has demonstrated to be useful. The qualitative study enabled identification of the most important concerns to the farmers related to the irrigation service and an idea on their satisfaction. It will serve the management of the irrigation (executive committee of the WUA) and agencies to best identify areas of action for a better and equitable service delivery.

Utility evaluation indicated that the physical infrastructure for irrigation alone cannot be successful unless an equally important aspect of the asset management and institutional capacity to handle are taken care of. It showed that in these schemes irrigation infrastructure improvement performs well if it is initiated by the water users themselves and if the required reforms for its management are well put in place. Otherwise, the farmers preferred the existing traditional and indigenous ways for managing the water. At Golgota Scheme, there has been no major improvement of the physical infrastructure since construction. The irrigation service, however, remained healthy. On the other hand, at Wedecha Scheme, the water diversion and distribution system was improved and several flow control structures were installed. Although it was intended to improve the services, farmers have damaged most of the flow control structures and their gates. The irrigation services were also unsatisfactory to the farmers regardless of the improvements.

Although the improvements in these community managed schemes are far to be named 'modernization', its concept still holds true. Facon and Renault (1998) view modernization as a technical and managerial upgrading and institutional reforms in irrigation schemes. When the schemes have been physically improved, it becomes incompatible with the existing practices in many cases of Ethiopia. There have been several small-scale irrigation schemes in Ethiopia in which the physical infrastructures have been improved recently. The situation is likely to be the case in these schemes as well. Therefore, physical irrigation infrastructure development in these schemes would have to always be combined with water management upgrading, capacity building and asset management. Utmost attention would also have to be put to the water users' views and their sense of ownership needs to be built.

10 EVALUATION AND THE WAY FORWARD

10.1 Prospects of the Awash River Basin

The Awash River Basin is the most utilized river basin of Ethiopia for irrigation developments. It is predominantly characterized by semi-arid and arid climates with huge evaporation loss. Its agro-climatic and demographic conditions make Awash River Basin a basin of great socio-economic significance in Ethiopia than any other. Modern irrigation developments started in this basin in the 1950's. The basin is particularly suitable for large and medium scale irrigation developments due to vast availability of irrigable land. Out of its irrigation potential of about 135,000 ha, the area currently under irrigation is about 35,000 ha, by large and medium scale schemes. There are also large-scale irrigation developments currently underway for an area of over 60,000 ha, which will be completed in less than three years time. Small-scale irrigation has been less practised in the basin due to the nature of the majority of the inhabitants moving from place to place. However, due to the recent government policy to settle pastoralists, small-scale irrigation has started to expand as well. As such, the vast pastoralist community in the basin has been transformed to semi-pastoralist thereby practising fixed farming activity.

Moreover, a recent master plan study of the basin has identified several sites for new large, medium and small-scale irrigation developments. It is likely that this huge expansion of irrigation in the basin can utilize the full potential of the basin in a decade time period. Awash River is also a source of water for all needs including municipal and large cattle population. The sectoral water competition in the basin is escalating. Extreme climatic conditions in the basin causing huge evaporation losses aggregated by the effects of climate change add up on the water stresses. The overall trend shows that Awash River Basin would shortly transform to a water stressed basin without adequate water management paradigm shift. To this end, there is a need for a basin wide re-institutionalization of water management and improvement of efficiency of water use particularly in agriculture.

10.2 Performance in large-scale irrigation schemes

With expected huge expansion of irrigation ranging from small-scale to large-scale in the next decade, water scarcity will be one of the top challenges that would be faced. This calls for a more efficient and sustainable use of the water resources. One of the major issues to be addressed in this regard is improving the performance of the existing irrigation schemes. In the existing large-scale irrigation schemes, there were no major concerns so far for improving irrigation performance and saving water. These schemes including Wonji-Shoa and Metahara have been operational for over 45 years and still use the same water management practices without major improvements. Annual water diversions to some of these schemes reach as high as 200%. However, the water abstraction patterns in these schemes cannot continue the same way as in the past for ever increasing water needs in the basin being added upon by natural causes for water scarcity.

Irrigation water conveyance, distribution and application at Wonji-Shoa and Metahara schemes are by gravity. Labour intensive manual gates have been in use for the overall flow control and regulation. The operation rules for these schemes have been in use for over 45 years with little or no major modifications. Manual operation of these schemes is not only laborious but has also resulted in an inappropriate supply of water to the schemes. Operators generally operate the structures based on calibrations done several years ago without any checks on the calibration. Recalibration of structures and regular flow monitoring is practically absent. The two schemes being located on the bank of the Main Awash River have access to excess water delivery. There were no major efforts so far at the schemes to match supplies with field demands to save water. There were also weak regulatory mechanisms by the Awash River Basin Authority so far regarding water diversions. The irrigation water fee paid by these schemes to the Authority is about 0.002 US$/m^3, which is extremely low to act as an incentive for saving water. In addition to excess water diversions, manual operation of structures has been associated with significant unsteadiness due to canal filling and emptying. The unsteady flow hydrodynamics were less understood both by operators and managers, and hence was a cause for significant operational losses and tail runoff.

Water is being distributed in these schemes by gravity through vast networks of earthen canals. Hydraulic performance at the main level (main, secondary and branch canals) is generally adequate. This is because due to a relatively larger and continuous flow at the main level, there is less need for maintenance due to less growth of aquatic plants. The scheduled annual maintenance in July and August is sufficient for the main system. There is little sedimentation in the main system of Wonji-Shoa Scheme, because water is supplied from a pumping station. But for Metahara Scheme, the sedimentation in the main level is huge. However, the annual maintenance is still sufficient to ensure adequate water conveyance in the main system. On the other hand, the sedimentation at lateral and sub-lateral (tertiary) levels appeared to totally silt up the canals occasionally. Particularly sedimentation of the sub-lateral (tertiary) canals critically affected the water delivery (hydraulic) performance at Metahara Scheme due to malfunctioning of sediment management devices and headwork structures.

Water delivery performance at both schemes has depicted that the proportion of excess water delivered to the fields was actually very small (less than 10% of excess). Regarding offtake deliveries in the head, middle and tail reaches, in the existing operation, tail offtakes deliver larger supplies at both schemes. At both schemes, head reach offtakes were supplied with nearly the right amount of water. At Metahara Scheme, middle offatkes were under-supplied, while at Wonji-Shoa, the delivery increased from the head to tail offtakes. The amount of water lost at the tail ends were more than 50% of the total surplus diversion for both schemes. Hydrodynamic simulation proved to be a useful tool for assessing the existing hydraulic performance and simulating alternative operations. With the DUFLOW simulations, it was learned that three key factors are of major significance for the hydraulic performance challenges in these schemes: inadequate manual operation; nature of offtake and water level control structures (sensitivity and flexibility); lag-time (hydraulic response time) of the systems.

Hydraulic performance and sustainability were found to be interlinked in the large-scale irrigation schemes. Sustainability in these schemes is manly a function of water use efficiency and hydraulic performance. Sustainability of these schemes can be affected in two ways as to their hydraulic performance. First, waterlogging and salinization are the main challenges to the sustainability due to excess and uncontrolled water diversion. In significant portions of these schemes, saline groundwater has risen to less than 1 m below the ground surface as a result of excess percolation particularly in the downstream ends of the schemes. Drainage and operational losses within the distribution system and excess on-farm applications also contribute their part to the rising groundwater levels. In this regard, water needs to be used efficiently and water diversions need to be well controlled. Secondly, sustainability also depends on water availability as the competing water needs in the basin have been increasing exponentially.

Huge irrigation expansion in the basin and other needs, including municipal and industrial, will limit water availability to these schemes and other existing schemes in the basin in the next few years. This in turn will call for a more efficient water management (water saving), without which the schemes would face significant water stresses which would challenge their sustainability. Water saving can be achieved by improving the hydraulic performance. Accordingly, enhancing hydraulic performance and hence water use efficiency in the large-scale schemes is crucial both under surplus and constrained water availability in order to ensure sustainability.

10.3   Performance in small community managed irrigation schemes

It is understood that small-scale irrigation is a dominant irrigation practice in Ethiopia for food production due to several factors including land ownership policy and demographics among others. These schemes have been playing a vital role in the efforts the country is making to ensure national food security. During the past two decades or so, small community managed schemes significantly changed a long existing situation of complete dependence on rainfall to a more intensive irrigated agriculture for vast farming communities. These not only ensured household food security of the smallholder farmers, but also enabled alleviation of poverty for the vast rural poor. Golgota and Wedecha community managed schemes have also greatly contributed in this regard.

However, the success of the community managed schemes of this study depends on a number of success factors. These factors include: institutional setups for effective management; capacity of the water management institutions; reliability of the water supply; sound irrigation fee policy for financial self sufficiency for operation, maintenance, management, etc. Reliability and flexibility of the irrigation supply are the main aspect farmers would like to have. Farmers of these schemes generally preferred to manage their water by themselves. This is so, because it gives them more flexibility for accessing irrigation water. However, self-management of these schemes was also associated with some flaws particularly in saving irrigation water and asset management. It was recognized that this is appropriate for schemes with relatively abundant water availability. In addition to routine maintenance fees, additional appropriate irrigation water fee will serve as a water saving incentive. Participatory management (irrigation agency and water users), while it proved to be useful in certain areas of management, it is not generally preferred by the water users of these schemes and most probably also of other schemes in the day-to-day management of irrigation water. However, owing to the low implementation capacity of the WUA in the majority of the schemes, entire irrigation management transfer in these schemes is not appropriate in many cases.

In Ethiopia, more than 90% of the total cultivated land is under smallholder farmers (rainfed and irrigated). Of the total smallholder agriculture, only nearly 10% has been irrigated so far. The rate of explanation in irrigated land area under small-scale schemes, however, has been increasing every year during the past decade. Productivity of irrigated smallholder farms is as high as 400% of the rainfed farms. Land and water availability in most of the community managed irrigation schemes in Ethiopia are constrained. The sources of water for these schemes comprise of mainly modern and traditional river diversions and micro reservoirs. Water scarcity in these schemes is either due to non-sustainability of the sources or poor infrastructure for water acquisition and distribution. The scale of water scarcity is, however different for different schemes. In general in several of these schemes limitation of water availability aggravated by climate change and expansion of irrigated land is expected to be intensified in the next few decades. Landholding sizes are generally extremely small. The situation with respect to land and water in the community managed schemes call for increasing productivity for both land and water.

Golgota and Wedecha irrigation schemes made significant contributions to the scheme beneficiaries in terms of food security and improving livelihoods. Land productivity in these schemes has been as high as 300% of the surrounding rainfed agriculture. However, there was a continuous shrinking in landholding sizes in these schemes, which requires a more steady increase in land productivity. Increasing the intensity of irrigation is a practical means to improve land productivity. With an annual irrigation intensity of 200 to 250%, land productivity in these schemes is competently high compared to similar schemes in Sub-Saharan Africa. Average water productivity (per diverted water) was 0.11 US$/ha for Golgota Scheme and 0.21 US$/ha Wedecha Scheme. These values are considered moderate compared with similar schemes in other regions of Ethiopia. However, the efficiency of water use in these schemes was extremely low. For Golgota Scheme, overall irrigation efficiency was about 20%, while that at Wedecha Scheme was about 40%. Owing to these low efficiencies, there remains a large potential to increase water productivity. Water saving incentives, enabling institutional setups, improvement of irrigation infrastructure, etc among others are the possible options. It was observed that irrigation water fee serves as an incentive for water saving. However, the fee needs to be reasonable and acceptable in order to ensure sustainability.

Less consideration is generally given to irrigation service quality in the small-scale irrigation schemes in Ethiopia. However, the success of these schemes greatly depends on the quality of the irrigation service. Often farmers' views on the condition of the services are different from the viewpoints of other stakeholders. In these schemes, irrigation flow measurements at offtake delivery points are considered irrelevant by several stakeholders. As such performance assessments in these schemes were often based on gross comparison of inputs and outputs (external performance). Such comparison, however provides little knowhow on the internal water management practices and variations in service levels among different groups of water users. Under data stress conditions for conventional data intensive internal irrigation performance indicators, assessment of irrigation service levels can be problematic. In the absence of quantitative data for computation of these indicators, assessment using fuzzy logic (fuzzy set theory) from qualitative data based on the water users' views can give a good idea on the internal performance and the bottlenecks. The methodology is not only relatively easier to apply, but also takes account of the farmers' perceptions, which are often overlooked.

10.4   Performance of existing schemes versus new irrigation developments

High variability of rainfall and occurrence of recurrent droughts and irregular dry spells are the main challenges to the vast Ethiopian rainfed agriculture. Though there has been acceleration in irrigation development during the last decade or so, the irrigation potential of Ethiopia is still largely unutilized. In view of the enormous potential and acute shortage of irrigation infrastructure in the country, there is indeed a need for even more accelerated development of irrigation in Ethiopia. With this understanding, the country had set out an ambitious irrigation development plans that would alleviate the immense dependence of the agriculture on rainfall. Over a double 5 year planning horizon from 2005 to 2014, it was planned to increase the irrigated land from the then

370,000 ha to 1.8 million ha (Federal Democratic Republic of Ethiopia (FDRE, 2009). Small-scale irrigation and rainwater harvesting schemes (RWH) account for about two-thirds of these expansions in these short-term plans for the following reasons (Awulachew, 2010): requirement of lower capital and technical investments, utilization of cheap and locally available labour, possibility to reach fragmented communities and smallholder households.

To this end, there has been a huge investment in irrigation development during the past decade. The total irrigated area has expended roughly by more than 100% since 2005. Moreover, there are several large-scale irrigation developments underway that are nearing completion. Although the progress in small-scale irrigation development (mainly diversion schemes) is a little behind schedule as per the plan, the expansion rate is still high. Rainwater harvesting has also been considered to be a viable solution for overcoming dry spells and ensuring household food security. All the irrigation expansion in Ethiopia is relevant in view of the development plans of the country. However, the performance of the existing irrigation schemes would also need to be considered utmost to ensure sustainability in the sector.

Irrigation performance assessment has not been well integrated and addressed in policy documents and operational plans of irrigation schemes in Ethiopia. Several of the completed schemes lack appropriate mechanisms for sustainable operation, maintenance and management. These include inadequate institutional setup, lack of clear operation and maintenance guidelines, lack of mechanisms for financial self sufficiency, inadequate asset management, lack of adequate irrigation scheduling, lack of evaluation of performance and monitoring, etc. As such in many irrigation schemes, several issues challenging the schemes satiability pose up in a few years of operation. Although there are schemes with good records of overall sustainability, the rate of mal-functionality is generally high, particularly in the medium and small-scale irrigation schemes. These schemes are generally handed over for routine operation and maintenance to the water users, who have little experience with modern irrigation management. Although there have been some irrigation performance assessment interventions, they have not been well tuned towards ensuring sustainability and benchmarking of good practices. Rather, the performance assessments largely focus on individual schemes and the results were less communicated to the stakeholders and were less implemented. Although farmers involvement in planning and development has been increasing with positive results, service oriented irrigation performance evaluation was given the least attention. This means that the water users' views were less entertained to improve the services once the irrigation schemes became operational. Both benchmarking performance and service oriented performance would need to be well integrated in the operational plans for overall sustainability.

Performance challenges in the large-scale public irrigation schemes are certainly different from the case in community managed schemes as has already been elaborated. In these schemes, service oriented performance evaluation is not relevant. Performance mainly relates to hydraulic aspects and efficiency of water use. Although there are several large-scale irrigation developments for community underway, existing large-scale irrigation schemes in Ethiopia were essentially public. These are operated under central scheme management and are meant for irrigation of industrial crops such as sugarcane, cotton and orchards. The irrigation infrastructure in these schemes is relatively sound. However, irrigation performance on supply-demand assessment, improving efficiency and hydraulic performance were not performed to the extent to make significant changes in water management in these schemes.

10.5   Irrigation performance and food security in the national context

Through various agricultural development and water management interventions, Ethiopia is currently on the verge of becoming food self sufficient. Immense efforts have been made in terms of agricultural water management to decrease the vulnerability of agriculture to temporal rainfall variability. Water management technologies ranging from in-situ soil moisture management to larger storage and diversion schemes have substantially reduced the susceptibility of crop production in unlikely situation of short or medium-scale rainfall failures. On the other hand, the sources of water particularly for small-scale irrigation schemes lack sound hardware and software. For 70% of the irrigated land in Ethiopia under small-scale irrigation, the country's efforts and plans to totally move away from food insecurity highly depend on the sustainability of these schemes. Currently, irrigated agriculture provides less than 15% of the total agricultural produce in Ethiopia. Apparently the contribution of irrigation to production would have to increase in order to further reduce vulnerability. A key factor to be addressed in this regard is to increase productivity and enhance the overall performance of the existing schemes. Expansion of irrigated land, while relevant, cannot be the only solution for improving production and ensuring the increasing population are fed. Ensuring sustainable performance and hence productivity of the existing schemes is equally vital. Approaches include intensifying irrigation, improving the physical infrastructure to increase command areas, implementing performance monitoring and remedial measures, improving the institutional setups and irrigation service delivery, etc.

The performance of large-scale irrigation schemes is related to the performance of small-scale schemes in the context of overall river basin water management. Fluctuations in the water sources (river flows) for the irrigation schemes are being aggravated by the impacts of climate change. Large-scale schemes, though have a relatively sound irrigation infrastructure, under the existing practices waste tremendous amount of water. These schemes with further expansion would undermine the success of small-scale schemes by limiting water availability. Significant water saving from these schemes consuming up to 50% more water of their actual demand has to be achieved. Thus, improving the overall efficiency of water use in the existing schemes (particularly large-scale schemes) is sought for their water supply sustainability as well as for sustainability of the small-scale schemes. Underlying factors for increasing agricultural production from irrigation schemes, such as increasing land productivity, expansion of command areas, etc. largely depend on sustainable scheme performance and appropriate irrigation service provision. Absence of irrigation performance and irrigation service monitoring leads to non-sustainability, contraction of irrigated areas, reduced productivity, etc., which will be later reflected in a larger picture of national food insecurity.

10.6   Conclusions and the way forward

In manually operated large-scale irrigation schemes, inadequate operation of water intake structures, offtakes, water levels control structures and night storage reservoirs largely contribute to low hydraulic performance. While the condition of the physical infrastructure is decisive for a good hydraulic performance, lack of adequate operation rules for these schemes plays an important role. The classical assumption by irrigation managers that head offtakes are always supplied with excess water does not always hold. The nature of flow control structures and the operation rules are the main factors determining the condition of the water delivery to offtakes at different reaches. In systems where discharges of offtake structures are more sensitive to water level variations than water level (cross) regulators, it is likely that tail and middle offtakes are supplied with excess. Typical such structures are adjustable overflow weirs as offtake structures and sluice gates as water level regulators as were depicted for Wonji-Shoa and Metahara schemes. In such manually operated schemes, substantial volumes of water would run to the downstream ends of the schemes, not only wasting considerable amounts of water but also threatening the sustainability, particularly for schemes in semi-arid regions. While hydraulic performance of such schemes can be evaluated from routinely monitored flow data, this would not explicitly indicate the causes for inadequate performance. Adequately calibrated hydraulic simulation model is proved to be a valuable tool for evaluation of the existing operation and for checking alternatives.

In community managed irrigation schemes, water users' evaluation of irrigation services can be different from the perceptions of other stakeholders. In schemes of smallholder farmers, the productivity of individual farmers and hence adequate internal irrigation service is more important than the overall efficiency of use of land and water resources of the scheme. For smallholder farmers (landholding sizes less than 1 ha), land productivity is directly proportional to landholding size. This is because farmers are willing to spend full time working on their farming as long as they are able to feed their families from their plot of land. Otherwise farmers look for a supplementary job, reducing their attention to their farming and productivity. In these schemes, appropriate institutional arrangements for water management essentially depend on the condition of the water source; i.e. on the degree of scarcity of water. Water management exclusively by the water users most likely would result in higher land productivity, but at an expense of lower water productivity and vice versa. In community managed irrigation schemes in least developed countries, where adequate data on water deliveries is not available, irrigation service levels can be well evaluated from qualitative and linguistic expression of the water users's perceptions.

This research has identified some fundamental performance challenges with respect to large-scale and small-scale community managed schemes to be addressed. Key recommendations for large scale-schemes include: 1. the long existing practice of operation of headworks would need to be replaced with alternative rules that will better match supplies with demands, not only for saving water, but also to ensure sustainability; 2. for optimum hydraulic performance and reduction of tail runoff, steady (quasy steady) water levels in the main system is utmost important. The combined operation of night storage reservoirs, offtakes and water level regulators plays a vital role in this regard. Reservoirs and flow control structures would need to be operated in such a way that nearly steady water levels are attained in the main system during irrigation hours.

For the community managed schemes key recommendations are: 1. irrigation flow measurement was almost not considered in these schemes; however, flow measurement at least at main delivery points would need to be considered for a better and sustainable irrigation water management. Purposes include reasonable billing of irrigation water fee (volumetric assessment) instead of area based water fee assessment, decision making for appropriate water diversion and delivery, evaluation of irrigation service delivery, etc., 2. there is no single appropriate institutional setup for irrigation water management for small-scale schemes. For sustainable water management in these schemes and for the schemes to address the national objective of food security, government entities at various levels and WUA need to put a special concern, particularly on suitable institutional setups based on typologies of the schemes and hence on sustainable asset management (operation and maintenance).

11. REFERENCES

Allen, R. G., Pereira, L. S., Raes, D. &Smith, M. (1998). *Crop evapotranspiration - Guidelines for computing crop water requirements - FAO Irrigation and drainage paper 56.* Rome, Italy: FAO.

Ankum, P. (2002). *Flow Control in Irrigation Systems: UNESCO-IHE Lecture Notes LN0086/04/1.* Delft, the Netherlands: UNESCO-IHE Institute for Water Education.

Awulachew, S. B. (2010).Irrigation potential in Ethiopia: Constraints and opportunities for enhancing the system. Addis Ababa, Ethiopia: International Water Management Institute.

Awulachew, S. B., Merrey, D. J., Kamara, A. B., Koppen, B. V., Vries, F. P. d. &Boelee, E. (2005). Experiences and opportunities for promoting small–scale/micro irrigation and rainwater harvesting for food security in Ethiopia. *Working paper* 98(IWMI): Colombo, Sri Lanka.

Behnke, R. &Kerven, C. (2013). *Counting the costs: replacing pastoralism with irrigated agriculture in the Awash Valley, north-eastern Ethiopia.* London, United Kingdom International Institute of Environment and Development (IIED)

Berhe, F. T., Melesse, A. M., Hailu, D. &Sileshi, Y. (2013). MODSIM-based water allocation modeling of Awash River Basin, Ethiopia. *Catena* 109: 118-128.

Bos, M. G., Burton, M. A. &Molden, D. J. (2004).Irrigation and Drainage Performance Assessment: Practical Guidelines. Oxfordshire, United Kingdom: CABI Publishing.

Bos, M. G., Burton, M. A. &Molden, D. J. (2005).Irrigation and Drainage Performance Assessment: Practical Guidelines. Trowbridge, United Kingdom: Cromwell Press.

Bos, M. G., D.H. Murray-Rust, Merrey, D. J., Johnson, H. G. &Snellen, W. B. (1994). Methodologies for assessing performance of irrigation and drainage management. *Irrigation and Drainage Systems* 7: 231-261.

Burt, C. M. &Styles, S. W. (1998). *Modern Water Control and Management Practices in Irrigation: Impact on Performance.* ITRC Report 98-001: Irrigation Training and Research Center.

Burt, C. M. &Styles, S. W. (2004). Conceptualizing irrigation project modernization through benchmarking and the rapid appraisal process. *Irrigation and Drainage, DOI: 10.1002/ird.127* 53: 145-154

Cakmak, B., Beyribey, M., YE.Yildirim &Kodal, S. (2004). Benchmarking performance of irrigation schemes: A case study from Turkey. *Irrigation and Drainage Systems* (53): 155-163.

Darghouth, S. (2005).Modernizing public irrigation institutions: The top priority for the future of sustainable irrigation. In *Nineteenth ICID Congress: Keynote Address* Beijing, China: International Commission on Irrigation and Drainage.

Dejen, Z. A., Schultz, B. &Hayde, L. (2012). Comparative irrigation performance assessment in community-managed schemes in Ethiopia. *African Journal of Agricultural Research* 7(35) 4956-4970.

Dubois, D. &Prade, H. (1978). Operations on fuzzy numbers. *International Journal of Systems Science* 9:6: 613-626.

El-Awad, O. M. (1991).Multicriterion Approach to the Evaluation of Irrigation Systems Performance. Vol. PhD thesisNewcastle, United Kingdom: University of Newcastle

Facon, T. &Renault, D. (1998).Modernization of irrigation system operations. In *Proceedings of the 5th ITIS network international meeting*Aurangabad, India: http://www.watercontrol.org.

Falkenmark, M., Lundquist, J. &Widstrand, C. (1989). Macro-scale water scarcity requires micro-scale approaches: Aspects of vulnerability in semi-arid development. *Nat. Resour. Forum* 13: 258-267.

FAO (2003).Agriculture, Food and Water, Natural Resources Management and Environment   Department, http://www.fao.org/DOCREP/006/Y4683E/Y4683E00.HTM. Rome, Italy: Food and Agriculture Organization of the United Nations

FAO (2006).Coping with water scarcity. Rome, Italy: Land and water development division.

FAO (2012).World Agriculture Towards 2030/2050: The 2012 revision. In *Agricultural Development Economics Division*Rome, Italy: Food and Agriculture Organization of the United Nations.

FAO (2013).Water Scrcity, Natural Resources and Environment Department Rome, Italy: Food and Agriculture Organization of the United Nations

FDRE (2009).Strategic Framework for Countrywide Water-Centered Development Program. Addis Ababa, Ethiopia

Gebreselassie, S. (2006).Land, Land Policy and Smallholder Agriculture in Ethiopia: Options and Scenarios. Brighton, United Kingdom: Institute of Development Studies.

Ghosh, S., Singh, R. &Kundu, D. K. (2005). Evaluation of Irrigation-Service Utility from the Perspective of Farmers. *Water Resources Management* 19: 467-482.

Girma, M. M. &Awulachew, S. B. (2007).Irrigation Practices in Ethiopia: Characteristics of Selected Irrigation Schemes. In *Working Paper 124*Colombo, Sri Lanka: International Water Management Institute

Hess, T. (2010).Water use: Water pressure. In *http://www.ecpa.eu/page/water-use*Brussels, Belgium.: European Crop Protection Association.

IFAD (2014).Water facts and figures, http://www.ifad.org/english/water/key.htm. Rome, Italy: International Fund for Agricultural Development.

Karady, G. (2001).Short Term Load Forecasting Using Neural networks and Fuzzy Logic. Arizona, United States: Arizona State University

Kazbekov, J., Abdullaev, I., Manthrithilake, H., Asad Qureshi a &Jumaboev, K. (2009). Evaluating planning and delivery performance of Water User Associations (WUAs) in Osh Province, Kyrgyzstan. *Agricultural Water Management* 96 1259-1267.

Kumar, P., Mishra, A., Raghuwanshi, N. S. &Singh, R. (2002). Application of unsteady flow inhydraulic model to a large and complex irrigation system. *Agricultural Water Management* 54: 49-66.

Malano, H., Burton, M. &Makin, I. (2004). Benchmarking performance in the irrigation and drainage sector: a tool for change. *Irrig. and Drain.* 53: 119-133.

Molden, D., Sakthivadivel, R., Perry, C. J., Fraiture, C. d. &Kloezen, W. H. (1998).Indicators for Comparing Performance of Irrigated Agricultural Systems. In *Research Report 20*Colombo, Sri Lanka: International Water Management Institute.

Molden, D. J. &Gates, T. K. (1990). Performance measures for evaluation of irrigation water delivery systems. *Journal of Irrigation and Drainage Engineering* 6(116): 804-823.

Moon, C., Lee, J. &Lim, S. (2007). A performance appraisal and promotion ranking system based on fuzzy logic: An implementation case in military organizations. *Applied Soft Computing* 10 (2010) 512-519.

Murray-Rust, D. H. &Snellen, W. B. (1993).Irrigation system performance assessment and diagnosis. Colombo, Sri Lanka: International Irrigation Management Institute.

Ngan, S.-C. (2011). Decision making with extended fuzzy linguistic computing, with applications to new product development and survey analysis. *Expert Systems with Applications* 38: 14052-14059.

Plusquellec, H. (2009). Modernization of large scale irrigation systems: Is it an achievable objective or a lost cause? *Irrigation and Drainage* 58: 104-120.

Renault, D. (2000). Aggregated hydraulic sensitivity indicators for irrigation system behavior. *Agricultural Water Management* 43 (2000): 151-171.

Renault, D. (2008).Service Oriented Management and Multiple Uses of Water in Modernizing Large Irrigation Systems. In *International symposium on multiple-use water services*Addis Ababa, Ethiopia: Food and Agricultural Organization of the United Nations

Renault, D., Facon, T. &Wahaj, R. (2007).Modernizing irrigation management – the MASSCOTE approach. Rome, Italy: Food and Agriculture Organization of the United Nations.

Renault, D. &Wahaj, R. (2007).Performance Indicators of Irrigation Service. Rome, Italy: FAO NRLW.

Sakthivadivel, R., Thiruvengadachari, S., Amerasinghe, U., Bastiaanssen, W. G. M. &Molden, D. (1999). *Performance Evaluation ofthe Bhakra Irrigation System, India, Using Remote Sensing and GIS Techniques.* Colombo, Sri Lanka: International Water Management Institute.

Sam-Amoah, L. K. &Gowing, J. W. (2001). Assessing the performance of irrigation schemes with minimum data on water deliveries. *Irrigation and Drainage* 50: 31-39.

Schultz, B. (2012).Land and Water Developemnt: Finding a blanance between planning, implementation, management and sustainability Delft, The netherlands.: UNESCO-IHE Institute for Water Education.

Schultz, B., Thatte, C. D. &Labhsetwar, V. K. (2005). Irrigation and drainage: main contributors to global food production. *Irrigation and Drainnage* 54: 263-278.

Şener, M., Yüksel, A. N. &Konukcu, F. (2007). Evaluation of Hayrabolu Irrigation Scheme in Turkey Using Comparetive Performance Indicators. *Journal of Tekirdag Agricultural Faculty* 4(1).

Shahrokhnia, M. A. &Javan, M. (2005). Performance assessment of Doroodzan irrigation network by steady state hydraulic modeling. *Irrigation and Drainage Systems* 19: 189-206.

STOWA (2004).Duflow Modelling Studio, Duflow Manual. The Netherlands.: STOWA/ MX.Systems.

Svendsen, M. &Small, L. E. (1990). Farmer's perspective on irrigation performance. *Irrigation and Drainage Systems* 4: 385-402.

Swennenhuis, J. (2006).CROPWAT for Windows Version 8.0 Rome, Italy: Water Resources Development and Management Service, FAO.

Swennenhuis, J. (2010a).CROPWAT 8.0: Water Resources Development and Management Service Rome, Italy: FAO.

Swennenhuis, J. (2010b).FAO CROPWAT 8.0, Water Resources Development and Management Service Rome, Italy.

Taddese, G., Sonder, K. &Peden, D. (2007).The Water of the Awash River Basin: A Future Challenge to Ethiopia. Addis Ababa, Ethiopia: ILRI.

Taddese, G., Sonder, K. &Peden, D. (2010).The water of the Awash River basin: A future challenge to Ethiopia. Addis Ababa, Ethiopia ILRI.

Tafesse, M. (2003).Small-scale irrigation for food security in Sub-Saharan Africa, CTA Working Document Number 8031. Addis Ababa, Ethiopia: The ACP-EU Technical Centre for Agricultural and Rural Cooperation.

Tanji, K. K. &Kielen, N. C. (2002). *Agricultural Drainage Water Management in Arid and Semi-Arid Areas: FAO Irrigation and Drainage Paper 61* Rome, Italy: FAO.

Tariq, J. A., Khan, M. J. &Kakar, M. J. (2004). Irrigation System Performance Monitoring as Diagnostic Tool to Operation: Case Study of Shahibala Minor of Warsak Gravity Canal. *Pakistan Journal of Water Resources* 8(1): 13-22.

Tariq, J. A. &Latif, M. (2010). Improving Operational Performance of Farmers Managed Distributary Canal using SIC Hydraulic Model. *Water Resour Management, 10.1007/s11269-010-9596-x.*

Unal, H. B., Asik, S., Avci, M., Yasar, S. &Akkuzu, E. (2004). Performance of water delivery system at tertiary canal level: a case study of the Menemen Left Bank Irrigation System, Gediz Basin, Turkey. *Agricultural Water Management* 65: 155-171.

USBR (2014). Parshall flumes-United States Bureau of Reclamation. http://www.usbr.gov/pmts/hydraulics_lab/pubs/wmm/chap08_10.html.

Vos, J. (2005). Understanding water delivery performance in a large-scale irrigation system in Peru. *Irrigation and Drainage* 54: 67-78.

WorldBank (2012).Ethiopia In *http://data.worldbank.org/country/ethiopia*Washington, DC, United States: The World Bank

Yercan, M., Dorsan, F. &Ul, M. A. (2004). Comparative analysis of performance criteria in irrigation schemes: a case study of Gediz river basin in Turkey. *Agricultural Water Management* 66 259-266.

Zadeh, L. A. (1965). Fuzzy Sets, Reprinted with permission from Information and Control, 8(3), 338-353. *Academic Press Inc. United States.*

Zimmermann, H. J. (2010).Fuzzy set theory: Adavanced Review. Vol. Volume 2Hoboken, NJ, United States: John Wiley & Sons, Inc. WIREs Comp Stat 2010 2 317–332.

ANNEXES

Annex A. List of symbols

Symbol	Description	Unit
ΔH	Change in water level	m
ΔQ	Change in discharge in parent canal	m³/s
Δq	Change in offtake discharge	m³/s
μ_i	Support of a fuzzy set element i	
A	Cross Sectional Area	m²
B	Width of a structure	m
C	Chezy coefficient	M¹ᐟ²/s
C_d	Discharge coefficient	-
F	Hydraulic flexibility	-
G	Acceleration due to gravity	m/s²
H	Head over the sill	m
H	Stage if water	m
h_o	Stage at which discharge is zero in a canal	m
K	Constant in stage-discharge relation equation	
M	Exponent in the stage-discharge relation equation	
M_i	Measured values	m or m³/s
N	Number of elements of a set	
p_A	Point adequacy inductor	ratio
P_A	Adequacy indicator	
P_D	Dependability indicator	
P_E	Equity indicator	
p_F	Point efficiency indicator	ratio
P_F	Efficiency indictor	
Q	Discharge	m³/s
Q	Discharge per unit width	m²/s
Q	Offtake discharge	m³/s
Q_D	Delivered volume	m³ or m³/s
Q_i	Intended discharge	m³/s
Q_R	Required volume	m³ or m³/s
R	Region	
S_f	Energy slope	m/m
S_i	Simulated values	m or m³/s
S_o	Bed slope	m/m
T	Time	s
U	Universal set	
V	Velocity	m/s
$\tilde{V}$	Depth averaged velocity	m/s
V_e	Effective volume	m³
V_i	Intended volume	m³
V_s	Supplied volume	m³
W	Width between two adjacent verticals in current metering	m
X	Longitudinal distance	m
Y	Water depth	m
M	Mean of measured values	m or m³/s

Annex B. Acronyms

ABA	Awash Basin Authority
AIDUIA	Annual Irrigation Water Delivery per Unit Irrigated Cropped Area
ARIS	Annual Relative Irrigation Supply
ARWS	Annual Relative Water Supply
B	Bad
BCM	Billion Cubic Meters
CANALMAN	Canal Management Model
CARIMA	Controlled Auto-Regressive Moving Average
CRM	Coefficient of Residual Mass
CROPWAT	Crop Water Requirement Model
CV	Coefficient of variation
CV_R	Spatial coefficient of variation
CV_T	Temporal coefficient of variation
DPR	Delivery performance ratio
DUFLOW	Dutch Flow Programme
EC	Electrical conductivity
ECe	Electrical Conductivity of soil extract
ET	Evapotranspiration
ETB	Ethiopian Birr
FAO	Food and Agricultural Organization of the United Nations
FDRE	Federal Democratic Republic of Ethiopia
FU	Farmer utility
G	Good
GDP	Gross Domestic product
GDP	Gross Domestic Product
GPS	Geographical Information System
GS	Golgota Scheme
HEC-RAS	Hydrologic Engineering Centre – River Analysis System
Hr	Hour
HVA	Handels Vereniging Amsterdam
IFPRI	International Food Policy Research Institute
IWMI	International Water Management Institute
M	Meter
M	Medium
ME	Model Efficiency
MLB	More or Less Bad
MLG	More or Less Good
Mm^3	Million cubic metres
MoWR	Ministry of Water Resources
MS	Metahara Scheme
MSL	Mean Sea Level
NRMSE	Normalized Root Mean Square Error
OE	Operational Efficiency
OPUCA	Output Per Unit Command Area
OPUIA	Output Per Unit Irrigated Area
OPUID	Output Per Unit Irrigation Water Delivered
OPUIS	Output Per Unit Irrigation Water Supply Or Diverted
OPUWC	Output Per Unit Water Consumed

OPUWS	Output Per Unit Water Supply/Diverted
PR	Performance Ratio
RD	Relative Delivery
RMC5	Resrvoir Main Canal 5
RMSE	Root Mean Square Error
RWH	Rain Water Harvesting
S	Second
SIC	Simulation of Irrigation Canals
SSI	Small Scale Irrigation
STOWA	Foundation for Applied Water Research (Dutch)
US$	US Dollars
VB	Very Bad
VG	Very Good
V_{inB}	Ineffective volume at the beginning of offtake flow
V_{inE}	Ineffective volume at the end of offtake flow
V_{inT}	Total ineffective delivered volume at an offtake
WL	Water Level
WS	Wedecha Scheme
WSS	Wonji-Shoa Scheme
WUA	Water Users Association

Annex C. Fuzzy numbers and fuzzy set operations and definitions

Fuzzy number

A fuzzy number is a quantity whose value is imprecise or not exact unlike a single valued number. A fuzzy number is represented by a fuzzy subset of the real line whose highest membership values are clustered around a given real number called the mean value (Dubois and Prade, 1978). Triangular and trapezoidal fuzzy numbers are shown in Figure C.1

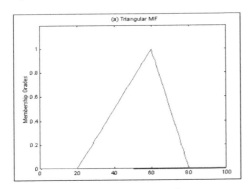
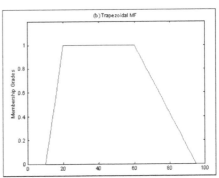

a. *Triangular fuzzy number*   b   *Trapezoidal fuzzy number*
Figure C.1. Representation of fuzzy numbers (Karady, 2001)

Union

The membership function of the Union of two fuzzy sets A and B with membership functions μ_A and μ_B respectively is defined as the maximum of the two individual membership functions as shown in Figure C.2. This is called the maximum criterion and is written as:

$$\mu_{A \cup B} = \max(\mu_A, \mu_B)$$
(C.1)

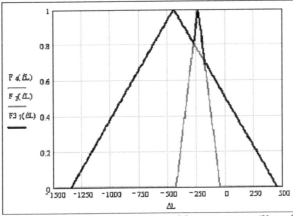

Figure C.2. Representation for a union of fuzzy two sets (Karady, 2001)

Intersection

The membership function of the Intersection of two fuzzy sets A and B with membership functions μ_A and μ_B respectively is defined as the minimum of the two individual membership functions as shown in Figure C.3. This is called the minimum criterion and is written as:

$$\mu_{A \cap B} = \min(\mu_A, \mu_B) \tag{C.2}$$

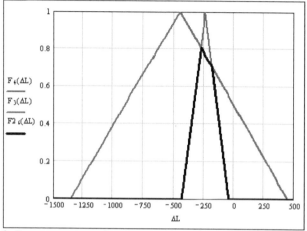

Figure C.3. Representation of an intersection of two fuzzy sets (Karady, 2001)

Normalized fuzzy sets

A fuzzy set is normalized when at least one of its elements have the maximum possible membership. In a membership range [0, 1], at least one element must have a membership of 1 for the fuzzy set to be considered normalized. Fuzzy sets can be normalized by dividing the membership grade of each element by the maximum support.

Convexity of fuzzy sets

A fuzzy set A is convex if and only if:

$$f_A[\lambda x_1 + (1 - \lambda)x_2] \geq Min\ [f_A(x_1), f_A(x_2)]; \text{ for all } x_1 \text{ and } x_2 \text{ in } X \text{ and all } \lambda \text{ in } [0, 1] \quad (C3)$$

It is described in Figure C.4.

Empty fuzzy sets

A fuzzy set is empty if and only if its membership function is zero on a Universe X.

Equal fuzzy sets

Two fuzzy sets A and B are equal, written as $A = B$, if and only if $fA(x) = fB(x)$ for all x in X. It shall be more easily written as $fA = fB$.

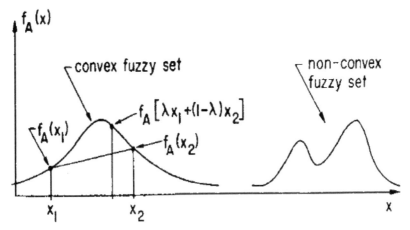

Figure C.4. Convex and non-convex fuzzy sets

Annex D. Discretization of unsteady flow equations

The unsteady flow equations (Saint-Venant equations) describe the hydrodynamics for open channel flow in time and space. These equations, which are the mathematical translation of the laws of conservation of mass and of momentum are:

$$\frac{\partial B}{\partial t} + \frac{\partial Q}{\partial x} = 0$$

(D.1)

$$\frac{\partial Q}{\partial t} + gA\frac{\partial H}{\partial x} + \frac{\partial(\alpha Q v)}{\partial x} + \frac{g|Q|Q}{C^2 AR} = a\gamma w^2 \cos(\Phi - \phi)$$

(D.2)

Where,

t	time [s]
x	distance as measured along the channel axis [m]
$H(x, t)$	water level with respect to reference level [m]
$v(x, t)$	mean velocity (averaged over the cross-sectional area) [m/s]
$Q(x, t)$	discharge at location x and at time t [m³/s]
$R(x, H)$	hydraulic radius of cross-section [m]
$a(x, H)$	cross-sectional flow width [m]
$A(x, H)$	cross-sectional flow area [m²]
$b(x, H)$	cross-sectional storage width [m]
$B(x, H)$	cross-sectional storage area [m²]
g	acceleration due to gravity [m/s²]
$C(x, H)$	coefficient of De Chézy [m$^{1/2}$/s]
$w(t)$	wind velocity [m/s]
$F(t)$	wind direction in degrees [degrees]
$f(x)$	direction of channel axis in degrees, measured clockwise from the north [degrees]
$g(x)$	wind conversion coefficient [-]
a	correction factor for non-uniformity of the velocity distribution in the advection term.

For a solution in DUFLOW, the unsteady flow equations are discredited in space and time using the four-point implicit Preissmann scheme. Defining a section Dx_i from node x_i to node x_{i+1} and a time interval Δt from time $t = t^n$ to time $t = t^{n+1}$, the discretization of the water level H can be expressed as (reference):

$$H_i^{n+\theta} = (1-\theta)H_i^n + \theta H_i^{n+1}$$

(D.3)

at node x_i and time $t + \theta\Delta t$ and

$$H_{i+1/2}^n = \frac{H_{i+1}^n + H_i^n}{2}$$

(D.4)

in between nodes xi and x_{i+1} at time t.

In a similar way other dependent variables can be approached. The transformed partial differential equations can be written as a system of algebraic equations by replacing the derivatives by finite difference expressions. These expressions approximate the derivatives at the point of references $(x_{i+1/2}, t^{n+\theta})$ as shown in Figure D.1.

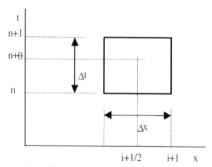

Figure D.1. The Four-Point Preissmann scheme

With initially:

$$H^0_{i-1/2} = H^n_{i+1/2}$$

$$B^n_{i+1/2} = B_{i+1/2}\left(H^n_{i+1/2}\right)$$

$$b^n_{i+1/2} = b_{i-1/2}\left(H^n_{i-1/2}\right)$$

$$B^{n,*}_{i+1/2} = B^n_{i+1/2} - b^n_{i+1/2}H^n_{i-1/2}$$

(D.5)

The equations for the conservation of mass and momentum are transformed into:

$$\frac{B^{*,n+1}_{i-1/2} + b^{n+1}_{i+1/2}H^{n+1}_{i-1/2} - B^n_{i+1/2}}{\Delta t} + \frac{Q^{n-\theta}_{i+1} - Q^{n-\theta}_{i}}{\Delta x_i} = 0$$

(D.6)

and

$$\frac{Q^{n+1}_{i+1/2} - Q^n_{i+1/2}}{\Delta t} + \frac{gA^*_{i+1/2}\left(H^{n+\theta}_{i-1} - H^{n+\theta}_i\right)}{\Delta x_i} + \frac{\alpha\left(\dfrac{Q^n_{i-1}}{A^*_{i-1}}Q^{n-1}_{i-1} - \dfrac{Q^n_i}{A^*_i}Q^{n-1}_i\right)}{\Delta x_i} +$$

$$+\frac{g\left(Q^{n-1}_{i+1/2}\left|Q^n_{i+1/2}\right|\right)}{\left(C^2AR\right)^*_{i-1/2}} = a^n\gamma\ (w^{n-1}_{i-1/2})\cos(\Phi^{n+1} - \phi)$$

(D.7)

    A mass conservative scheme for water movement is essential for proper water quality simulation. If the continuity equation is not properly taken into account, the calculated concentration will not match the actual concentration. The mass conservative scheme is based on the fact that the error made in the continuity equation will be corrected in the next time step. Mass conservation is therefore guaranteed. The * (like in $A^*i+1/2$) expresses that these values are approximated at time $t^{n+\theta}$.

    This discretization is of second order in time and place if the value $\theta = 0.5$ and it can be shown that in this case the discredited system is mass-conservative. In most applications, a somewhat larger θ value, such as 0.55 is used in order to obtain a better stability (Roache, 1972). The values indicated with (*) are computed using an iterative process. For example, a first approximation of A is $A^* = A^n$, which is adjusted in subsequent iteration steps:

$$A^* = \frac{\left(A^n + A^{n-1,*}\right)}{2}$$

(D.8)

Where $A^{n+1,*}$ is the new computed value of A^{n+1}.

So finally, for all channel sections in the network two equations are formed, which have Q and H as unknowns on the new time level t^{n+1}:

$$Q_i^{n+1} = N_{11}H_i^{n+1} + N_{12}H_{i+1}^{n+1} + N_{13}$$

(D.9)

$$Q_{i+1}^{n+1} = N_{21}H_i^{n+1} + N_{22}H_{i+1}^{n+1} + N_{23}$$

(D.10)

Annex E. Guidelines for calculation of crop water requirements

Crop evapotranspiration (ETc) is the evapotranspiration from disease-free, well-fertilized crops, grown in large fields, under optimum soil water conditions, and achieving full production under the given climatic conditions (Allen et al., 1998). Crop water requirement may simply be defined as the amount of water required by a crop for its development and maturity. Although crop evapotranspiration can be calculated from climatic data and by integrating directly the crop resistance, albedo and air resistance factors, better the Penman-Monteith method is used for the estimation of the standard reference crop and be used to determine crop evapotranspiration.

$$ET_c = K_c \, ET_o \tag{E.1}$$

Where Kc is crop coefficient and ET_o is reference evapotranspiration (mm or mm/day). Kc, the ratio ET_c/ET_o can be experimentally determined.

Reference evapotranspiration (ETo) is the rate of evapotranspiration from a reference crop (green grass) with a height of 8-15 cm, actively growing, completely shading the ground, and no short of water (Allen et al., 1998). The FAO Penman-Monteith method is selected as the method by which the ETo can be unambiguously determined, and the method which provides consistent ETo values in all regions and climates. The modified Penman-Monteith equation reads as:

$$ET_o = \frac{0.408\Delta(R_n - G) + \gamma \dfrac{900}{T+273} u_2(e_s - e_a)}{\Delta + \gamma(1 + 0.34u_2)} \tag{E.2}$$

Where:

ETo	reference evapotranspiration [mm day^{-1}]
Rn	net radiation at the crop surface [MJ m^{-2} day^{-1}]
G	soil heat flux density [MJ m^{-2} day^{-1}]
T	mean daily air temperature at 2 m height [°C]
u_2	wind speed at 2 m height [m s^{-1}]
e_s	saturation vapour pressure [kPa]
e_a	actual vapour pressure [kPa]
$e_s - e_a$	saturation vapour pressure deficit [kPa]
Δ	slope vapour pressure curve [kPa °C^{-1}]
g	psychrometric constant [kPa °C^{-1}].

FAO CROPWAT model, which was used for determination of crop water requirements in this study, uses the Modified Penman-Monteith equation. ETo is only dependent on climatic parameters. Meteorological data required for calculation of ETo include solar radiation, temperature, relative humidity, wind speed.

However, evapotranspiration (crop water requirement) depends on the following three factors.

Weather parameters

The principal weather parameters affecting evapotranspiration are radiation, air temperature, humidity and wind speed.

Crop factors

The crop type, variety and development stage should be considered when assessing the evapotranspiration from crops grown in large, well-managed fields. Differences in resistance to transpiration, crop height, crop roughness, reflection, ground cover and crop rooting characteristics result in different ET levels in different types of crops under identical environmental conditions.

Management and environmental conditions

Factors such as soil salinity, poor land fertility, limited application of fertilizers, the presence of hard or impenetrable soil horizons, the absence of control of diseases and pests and poor soil management may limit the crop development and reduce the evapotranspiration. Other factors to be considered when assessing ET are ground cover, plant density and the soil water content. The effect of soil water content on ET is conditioned primarily by the magnitude of the water deficit and the type of soil. On the other hand, too much water will result in waterlogging which might damage the root and limit root water uptake by inhibiting respiration (Allen et al., 1998).

Annex F. Parshall flume dimensions and constants

The Parshall flume is an open channel flow measuring structure originally developed to measure surface water and irrigation flows. The Parshall flume is now frequently used to measure industrial discharges, municipal sewer flows, and influent / effluent at wastewater treatment plants. Development of the Parshall flume began in 1915 by Dr. Ralph L. Parshall of the U.S. Soil Conservation Service. The drop in elevation through the throat of the flume produces supercritical flow. With supercritical flow, only one head measurement is necessary to determine the flow rate, greatly simplifying the use of the flume.

The design of the Parshall flume consists of a uniformly converging upstream section, a short parallel throat section (the width of which determines the flume size), and a uniformly diverging downstream section (Figure F.1). The floor of the flume is flat in the upstream section, slopes downward in the throat, and then rises in the downstream section; ending with a downstream elevation below that of the upstream elevation. Although the basic shape of all Parshall flumes is the same, the flumes are not scale models of each other, so that the discharge equation for each flume had to be obtained by direct calibration (USBR, 2014).

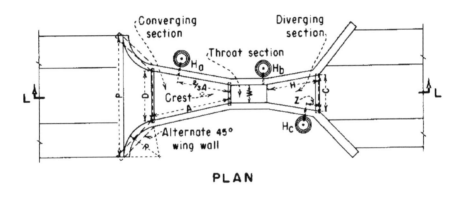

PLAN

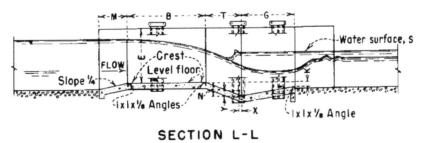

SECTION L-L

Figure F.1. Parshall flume dimensions (U.S. Natural Resources Conservation Services)

The size of a Parshall flume is designated by the size of throat, and the free flow (discharge) through it is given by:

$$Q = C * H^n \tag{F.1}$$

Where H is water depth measured in the converging (upstream) section, C and n are

constants to be determined for each flume by calibration (as shown in Table F.1). In the
community managed small scale irrigation schemes of this study, Parshall flumes were
used to measure relatively smaller discharges in canals.

Table F.1. Constants for Parshall Flume free flow equation

Throat width, W, (in, ft)	Throat width, W (cm, m)	C	n
1 in	2.5 cm	0.060	1.55
2	5.1	0.121	1.55
3	7.6	0.177	1.55
6	15.2	0.381	1.58
9	22.9	0.535	1.53
12	30.5	0.691	1.522
18	45.7	1.056	1.538
2 ft	610	1.429	1.550
3	914	2.184	1.566
4	1.219 m	2.954	1.578
5	1.524	3.732	1.587
6	1.829	4.518	1.595
7	2.134	5.313	1.601
8	2.438	6.115	1.607
10	3.048	7.463	1.6
12	3.658	8.859	1.6
15	4.572	10.96	1.6
20	6.096	14.45	1.6
25	7.620	17.94	1.6
30	9.144	21.44	1.6
40	12.19	28.43	1.6
50	15.24	35.41	1.6

Annex G. Stage-discharge relations for open channel flow

It is often not easy to measure open channel flows continuously, and hence a relationship between stages of water and discharge can be established at a control section to determine discharges for any observed stages. For open channel flow, there is a distinct relationship between the flow depth (stage) and discharge for a given channel characteristics. This relation is called a rating curve or Q-h relation. In the community managed schemes (Golgota), irrigation flows were measured using a Q-h relation at control sections in the canal. The discharge equation (rating curve) for flow in open channels is given by:

$$Q = a(h - h_o)^b \tag{G.1}$$

Where: Q is discharge (m³/s), h is measured water level (m), h_o is water level corresponding to $Q = 0$, and a and b are constants. h_o, the stage for which discharge is zero can be easily determined from the arithmetic plot between stage and discharge. In order to determine the coefficients a and b, the power equation G.1 can be transformed into a linear form by taking the logarithms of both sides:

$$\log Q = \log a + b \log(h - h_o) \tag{G.2}$$

From the straight line plot of date of $\log Q$ versus $\log (h - h_o)$, $\log a$ and b can be easily determined from a linear regression. At Golgota Scheme, for the rating curve established at the head of the canal (Figure G.1) for measuring irrigation water diversion, the rating curve equation was:

$$Q = 1.76 * (h - 0.05)^{0.95} \tag{G.3}$$

Where: Q is discharge (m³/s), h flow depth (m).

Figure G.1. a. Current metering, b. Flow depth measurement at a control section, for stage-discharge relation at Golgota Scheme

Annex H. Meteorological data at the irrigation schemes

Table H.1. Monthly meteorological data at Metahara Scheme

	Jan	Feb	Mar	Apr	May	Jun	Jul	Aug	Sep	Oct	Nov	Dec	Annual
Mean rainfall, mm	16	30	47	39	36	27	122	130	46	25	8	7	533
Mean max. T, °C	30.8	32	33.7	34.2	35.6	36.4	33.1	32.4	33.8	33	31.7	30.3	
Mean min. T, °C	14.8	16	18.4	19.1	19.1	21.3	20.3	19.8	19.1	16.2	14.1	13.8	
RH, %	63	60	59	59	55	50	58	61	59	57	61	62	
Wind speed, km/h	8.6	9.2	8.5	7.7	7.7	13.0	14.0	10.6	7.0	6.5	7.4	8.3	
Sunshine, hrs	8.0	9.0	8.7	8.0	8.5	7.5	7.0	7.0	8.0	8.0	8.5	8.8	

Table H.2. Monthly meteorological data at Wonj-Shoa Scheme

	Jan	Feb	Mar	Apr	May	Jun	Jul	Aug	Sep	Oct	Nov	Dec	Annual
Mean rainfall, mm	16	32	48	41	36	27	123	129	45	27	9	7	540
Mean max. T, °C	26.5	28.0	29.1	29.1	29.7	28.8	26.2	25.8	26.3	27.1	26.7	26.1	
Mean min. T, °C	13.3	14.5	15.5	15.9	16.3	17.4	16.9	16.9	16.0	13.1	12.2	11.9	
RH, %	52	48	50	54	52	56	64	67	64	52	49	51	
Wind speed, km/h	4.8	4.9	4.4	3.9	3.6	4.8	5.0	3.8	2.6	3.3	4.5	4.8	
Sunshine, hrs	8.7	8.9	8.3	8.4	8.3	8.4	6.9	7.5	7.3	8.9	9.6	9.4	

Table H.3. Monthly meteorological data at Golgota Scheme

	Jan	Feb	Mar	Apr	May	Jun	Jul	Aug	Sep	Oct	Nov	Dec	Annual
Mean rainfall, mm	35	12	57	41	26	32	139	140	46	39	5	11	585
Mean max. T, °C	29.2	31.4	32.4	33.0	34.9	34.8	31.7	30.9	32.1	31.9	29.9	28.9	
Mean min. T, °C	15.1	14.5	17.0	18.1	19.1	20.9	19.3	19.0	18.7	15.6	13.2	12.7	
RH, %	56	56	56	56	48	46	58	63	60	48	50	54	
Wind speed, km/h	5.1	5.6	5.5	5.6	6.5	10.0	10.5	8.3	5.8	5.0	5.1	5.1	
Sunshine, hrs	8.8	8.5	8.2	7.4	9.0	7.4	6.9	7.3	7.5	8.7	8.7	8.8	

Table H.4. Monthly meteorological data at Wedecha Scheme

	Jan	Feb	Mar	Apr	May	Jun	Jul	Aug	Sep	Oct	Nov	Dec	Annual
Mean rainfall, mm	10	28	49	57	51	90	211	197	93	21	10	3	822
Mean max. T, °C	26.7	28.3	28.9	28.5	29.1	27.9	24.6	24.8	25.7	26.1	25.7	26.5	
Mean min. T, °C	8.9	10.1	11.9	12.7	12.3	12.2	13.5	13.5	12.2	9.3	7.7	7.4	
RH, %	50	47	47	50	49	58	69	71	66	51	46	48	
Wind speed, km/h	7.0	8.1	8.2	8.1	8.0	5.2	4.6	5.3	4.5	7.4	8.0	8.2	
Sunshine, hrs	8.7	8.2	7.8	6.9	8.0	6.4	5.0	5.7	6.8	8.8	9.5	9.3	

Annex I. Samenvatting

De twee eindige hulpbronnen voor geïrrigeerde landbouw, land en water, krimpen wereldwijd. Om dezelfde reden, is het vergroten van het areaal geïrrigeerd land gedurende de laatste twee decennia sterk afgenomen. Met de beperkte voorraden aan zoetwater en land, en de steeds toenemende concurrentie met betrekking tot deze voorraden, is het nodig dat geïrrigeerde landbouw (grootste gebruiker van wereldwijde zoetwater voorraden) haar gebruik van deze middelen verbetert. Er is een grote overeenstemming dat het groeitempo van irrigatiewater onttrekking in de komende decennia zal vertragen. Daarom zou het grootste deel van de vereiste verhoging van de landbouwproductie moeten worden gerealiseerd op de al bestaande landbouwgronden. Dit zou mogelijk zijn door betere irrigatie en drainage praktijken, toename van irrigatie intensiteit, en verbeterde dienstverlening op het gebied van irrigatie.

Ethiopië is een van de minst ontwikkelde landen in de Hoorn van Afrika, met een totale oppervlakte van 1,13 miljoen km^2. De oppervlakte die geschikt is voor landbouw wordt geschat op 72 miljoen hectare, terwijl recente rapporten aangeven dat slechts ongeveer 25% (15 miljoen hectare) in cultuur is gebracht. Landbouw ondersteunt direct ongeveer 85% van de bevolking van Ethiopië en vormt meer dan 80% van de waarde van de export. De landbouwsector bleef echter tot voor kort onderontwikkeld en weinig productief. De sector wordt gedomineerd door kleinschalige landbouw voor eigen gebruik, uitdagingen in verband met slecht ontwikkelde kweekmethoden, bodemdegradatie en een grote afhankelijkheid van de regenval. De overgrote meerderheid (95%) van de Ethiopische landbouw is afhankelijk van de neerslag die een grote variatie in tijd en ruimte kent. In de afgelopen jaren heeft het herhaaldelijk optreden van regenval tekorten (droogten), verergerd door de gevolgen van de klimaatverandering, een aanzienlijke bevolking in de laaglanden in gevaar gebracht. De watervoorraden van Ethiopië zijn enorm; zij bedragen jaarlijks ongeveer 125 miljard m^3 aan oppervlaktewater potentieel en een geschat grondwater potentieel van 2,6 miljard m^3. Het totale irrigeerbare land potentieel wordt geschat op 5,3 miljoen ha, met inbegrip van oppervlaktewater, grondwater en het verzamelen van regenwater. Het geïrrigeerde gebied bedraagt tot op heden 700.000 ha, en omvat systemen van alle schalen. Dit getal laat zien dat het geïrrigeerde gebied slechts 12% van het potentieel beslaat en 5% van de oppervlakte landbouwgrond.

Ethiopië is momenteel begonnen aan een versnelde ontwikkeling van geïrrigeerde landbouw, waarbij de planning is dat het geïrrigeerde areaal in vijf jaar wordt verdrievoudigd. Blijkbaar is uitbreiding van het geïrrigeerd areaal door middel van nieuwe irrigatie ontwikkelingen relevant, gezien het onderbenutte potentieel. Zorgen voor duurzaamheid van de bestaande systemen is echter eveneens van vitaal belang, dat wordt duidelijk over het hoofd gezien. De meerderheid van de operationele irrigatieprojecten in het land worden gekenmerkt door een laag niveau van technische, hydraulische en operationele dienstverlening. De tekortkomingen betreffen onvoldoende roulatie van irrigatiewater voorziening, onvoldoende planning van het beheer, het ontbreken van adequate institutionele organisatie voor het beheer, ontoereikende fysieke faciliteiten op het gebied van waterbeheer, sedimentatie in de kanalen, gebrek aan degelijke meting van irrigatiewater, enz. Sommige van deze uitdagingen zijn cruciaal voor kleinschalige door de gemeenschap beheerde systemen, terwijl anderen fundamenteel zijn voor grootschalige systemen.

Dit proefschrift beschrijft en evalueert het functioneren van twee grootschalige en twee door de gemeenschap beheerde irrigatiesystemen in Ethiopië. De grootschalige systemen staan bekend als Wonji-Shoa en Metahara, terwijl Golgota en Wedecha de door de gemeenschap beheerde systemen zijn. Wonji-Shoa, en Metahara zijn ontwikkeld in de vijftiger en zestiger jaren van de twintigste eeuw en irrigeren gebieden van respectievelijk 6.000 en 11.500 ha. Ze bevinden zich in de vallei van de Awash rivier, in de centrale Rift vallei van Ethiopië, op ongeveer 100 km van elkaar met het Metahara systeem benedenstrooms. Dit zijn beide door de overheid beheerde systemen waar uitsluitend suikerriet wordt geteeld en de Awash rivier wordt gebruikt als bron voor irrigatiewater. Golgota is een van de door de gemeenschap beheerde systemen in deze studie, en is in hetzelfde stroomgebied gelegen tussen de Wonji-Shoa en Metahara systemen. Met een tijdelijke schanskorf wordt water aan dit systeem, met een nominale commando oppervlakte van 600 ha, geleverd. Het door de gemeenschap beheerde Wedecha systeem is gelegen in de centrale hooglanden van Ethiopië, ook in het stroomgebied van de Awash rivier, en heeft een nominale oppervlakte van 360 ha. Irrigatie water wordt geleverd uit een reservoir dat is ontstaan door een kleine dam in de Wedecha, een kleine zijrivier van de Awash rivier.

Dit promotie onderzoek was gericht op de evaluatie van het functioneren van de hydraulica en wateraanvoer in de grootschalige systemen met als doelstellingen de evaluatie van de bestaande regels voor waterbeheer en het voorstellen van alternatieve mogelijkheden voor een meer doeltreffend beheer, duurzaamheid en waterbesparing. Anderzijds betrof het onderzoek een vergelijkende evaluatie van het functioneren en de resultaten van de interne irrigatie dienstlevering in de twee door de gemeenschap beheerde systemen.

Het stroomgebied van de Awash rivier is het meest gebruikte stroomgebied in Ethiopië voor irrigatie. Het is een stroomgebied met een groot sociaaleconomisch belang, vanwege de route door de droogste noordoostelijke Rift vallei regio. De rivier is de enige bron van water voor meer dan 5 miljoen veehouders en deeltijd veehouders met hun vee in de regio. Bovendien, het is een bron voor de drinkwatervoorziening van verschillende steden langs de rivier. Tegenwoordig zijn in het stroomgebied een aantal grote en middelgrote irrigatie ontwikkelingen gaande. Bovendien zijn er grote aantallen door de gemeenschap beheerde irrigatiesystemen in aanbouw in een poging van de regering om de voedselzekerheid te verbeteren door middel van het transformeren van de grote plattelandgemeenschap in een deeltijd plattelandgemeenschap. Als zodanig is de concurrentie voor water in het stroomgebied in de afgelopen jaren geïntensiveerd en is er een toenemende druk op de bestaande systemen om het water efficiënter te gaan gebruiken. Uiteraard, zal de toenemende vraag naar water in het stroomgebied het water aandeel van de bestaande systemen reduceren. Dit vraagt vervolgens om een efficiënter irrigatie waterbeheer dat betere operationele efficiëntie, toereikendheid en gelijkheid waarborgt. De Wonji-Shoa en Metahara systemen behoren tot de belangrijkste irrigatiesystemen in het stroomgebied, die hun waterbeheer moeten analyseren en verbeteren. De door de gemeenschap beheerde systemen, zoals de twee systemen in deze studie, spelen een belangrijke rol voor de voedselzekerheid en het bestrijden van armoede op het platteland. Om de duurzaamheid van deze systemen te garanderen moeten de irrigatie dienstverlening, productiviteit van water en de institutionele aspecten voor het waterbeheer worden geanalyseerd en verbeterd.

Beoordeling van het functioneren van irrigatie en drainage betreft een systematische waarneming en interpretatie van het beheer van de systemen, met als doel ervoor te zorgen dat de inzet van middelen, het operationele beheer, de beoogde resultaten en de benodigde maatregelen conform de planning worden gerealiseerd. De algemene doelstelling van de evaluatie van het functioneren is om verbetering te garanderen. Beoordeling van het functioneren is tijdens de laatste twee decennia een uitgebreid onderwerp van studie en zorg geweest in het kader van de afnemende land en water voorraden en de noodzaak om de productiviteit van bestaande irrigatiesystemen te verhogen. Dienovereenkomstig, hebben een vrij groot aantal onderzoekers de verschillende aspecten van het functioneren van irrigatiesystemen in de wereld bestudeerd en voorstellen gedaan voor verbetering. Er zijn echter in het verleden, vooral betreffende grote en middelgrote systemen, bijna geen initiatieven voor evaluaties van het functioneren van irrigatiesystemen in Ethiopië geweest. In dit onderzoek zijn de gesignaleerde problemen bij het functioneren van de grootschalige systemen hydraulisch (water aanvoer), waterbesparing en daarmee samenhangende milieuaspecten, wateroverlast en verzilting. Anderzijds hebben de bij de door de gemeenschap beheerde systemen gesignaleerde problemen ten aanzien van het functioneren betrekking op de dienstverlening van irrigatiewater voorziening, productiviteit van land en water en institutionele aspecten van het waterbeheer.

De handmatige bediening van kunstwerken voor waterbeheer van de Wonji-Shoa en Metahara systemen is niet alleen arbeidsintensief en bewerkelijk, maar ook niet effectief. Het complexe hydrodynamische gedrag van de systemen wordt niet goed begrepen, en het bestaande beheer houdt weinig rekening met deze effecten. Het hydraulisch functioneren is eerst voor elk systeem geëvalueerd op basis van routinematig gemeten afvoeren bij inlaatwerken die zijn geclassificeerd als bovenstrooms, midden en benedenstrooms. Kanalen van 9 en 11 km lengte, met respectievelijk 16 en 15 inlaatwerken zijn geanalyseerd in de Wonji-Shoa en Metahara systemen. Afvoeren in de inlaatwerken zijn gedurende drie maanden (januari, februari en maart) voor de jaren 2012 en 2013 twee keer per dag gemeten. Dit zijn de maanden met lage afvoer in de Awash rivier, waarin de beschikbaarheid van water minimaal is. Toereikendheid (relatieve watervoorziening), watervoorziening aan de inlaatwerken, billijkheid en betrouwbaarheid zijn als indicatoren voor de betrouwbaarheid van de watervoorziening gebruikt. Bovendien is een vergelijking gemaakt van de jaarlijkse voorziening van irrigatiewater versus de vraag op basis van gemeten wateraanvoeren met behulp van niveau versus afvoer relaties bij de inlaatwerken van de belangrijkste kanalen.

De situatie met betrekking tot het functioneren van de wateraanvoer is op basis van routinematig verzamelde afvoergegevens van de twee grootschalige irrigatiesystemen duidelijk aangetoond. Het routinematig meten van afvoer gegevens is tijdrovend en omslachtig. Men kan echter de huidige omstandigheden observeren waardoor de resultaten betrouwbaarder zijn. Als eerste stap is de gemeten wateraanvoer aan de inlaat van de kanalen vergeleken met de berekende hoeveelheden, en voor beiden werd een significant te grote hoeveelheid gevonden. Vervolgens is voor geselecteerde (hoofd en secundaire) kanalen, de afvoer over de inlaatwerken gedurende twee jaar tijdens de drie droge maanden gevolgd om deze te gebruiken voor het bepalen van het functioneren van de wateraanvoer.

De jaarlijkse gemeten wateraanvoeren overschrijden de jaarlijkse vraag met 41 en 24% bij respectievelijk de Wonji-Shoa en Metahara systemen. De resultaten gaven aan dat in tegenstelling tot de klassieke veronderstelling dat de bovenstroomse inlaatwerken grotere hoeveelheden water leveren, voor beide systemen, de inlaatwerken in het benedenstroomse deel bij het bestaande beheer een overmaat en een grotere hoeveelheid water leverden. Het bleek dat de wateraanvoer bij de bovenstroomse inlaatwerken bij beide systemen acceptabel is, dit komt door redelijk kleine schommelingen in het waterpeil in de bovenstroomse sectie. Het legen en vullen van de kanalen had de ergste gevolgen voor het hydraulisch functioneren (adequaatheid) van de inlaatwerken in de midden sectie van het Metahara systeem. Voor het Wonji-Shoa systeem, vertoonde de wateraanvoer een afname van het bovenstroomse naar het benedenstroomse deel. Overtollige wateraanvoer in het benedenstroomse deel was te wijten aan het gebrekkige functioneren en het zeer gevoelige karakter van de kunstwerken. Al met al, was de wateraanvoer bij de inlaatwerken bij het huidige beheer voor beide systemen redelijk acceptabel, hoewel er aanzienlijke schommelingen van het ene jaar op het andere waren. Dit kwam omdat het gedeelte van het overtollig geleverde water dat verloren ging in de tertiaire eenheden bij beide systemen relatief klein was in vergelijking met de verliezen buiten de landbouwkavels (distributie), operationele verliezen op niveau van het hoofdsysteem en benedenstroomse afvoer. Metingen en efficiëntie indicatoren hebben aangegeven dat percolatie verliezen op veldniveau goed zijn voor slechts 20 en 10% van de totale jaarlijks aangevoerde hoeveelheid water bij respectievelijk de Wonji-Shoa en Metahara systemen. De resterende hoeveelheden percoleren vanuit de wateraanvoer kanalen en worden afgevoerd naar de open drains van de systemen en naar de verzilte benedenstroomse gedeelten.

Er kan worden waargenomen dat de operationele verliezen als gevolg van voortdurende schommelingen in de afvoer, de aard van de kunstwerken voor het waterbeheer en plotselinge sluiting van de benedenstroomse inlaatwerken er in resulteren dat enorme hoeveelheden overtollig water bij het afvoerpunt terecht komen. Als zodanig bleek uit de waarnemingen dat de aanvoer van irrigatie water aan het irrigatie blok genaamd 'Noord blok', benedenstrooms gelegen in het Metahara systeem, meer dan het dubbele van de vraag was. Aanvoer van irrigatie water aan de benedenstroomse sectie van het Wonji-Shoa systeem was ook bijna zo hoog als 200% van de vraag. Dit heeft vervolgens de duurzaamheid van deze systemen aan de kaak gesteld, waardoor ondiepe grondwaterstanden tot minder dan 1 meter onder het maaiveld ontstonden. Operationele beslissingen betreffende maatregelen voor het waterbeheer kunnen het functioneren van de wateraanvoer aanzienlijk verbeteren. Sommige van deze maatregelen op basis van waarnemingen in het veld, metingen en evaluatie worden aanbevolen in dit proefschrift.

Hydrodynamische simulatie modellen zijn handige hulpmiddelen om de complexe hydrodynamica van kanaal irrigatiesystemen te begrijpen en hun hydraulisch functioneren te evalueren. De effecten van verschillende operationele maatregelen op de hydrodynamica en het resulterende functioneren kunnen worden geëvalueerd. Deze modellen zijn gebruikt door verschillende onderzoekers voor de evaluatie van het functioneren van irrigatiesystemen of ter ondersteuning van het verbeteren van het beheer. De toepassing van deze modellen was tot nu toe echter vooral gericht op irrigatieprojecten die werden beheerd door individuele, of groepen van watergebruikers. In dit geval is een hydraulisch model gebruikt voor het irrigatiesysteem van de Metahara suikerriet plantage zonder individuele watergebruikers. DUFLOW, een eendimensionaal hydrodynamisch model, is gekalibreerd en gebruikt om het huidige functioneren van het systeem met betrekking tot de water aanvoer (hydraulica) te analyseren. Gemeten afvoeren van 16 inlaatwerken langs het kanaal en de gemeten waterpeilen op twee locaties (1+300 en 7+100) in het kanaal systeem zijn gebruikt voor de kalibratie. Het model is gevalideerd met gemeten afvoeren van de inlaatwerken bij andere hydrodynamische condities dan bij de kalibratie. Afvoeren voor de kalibratie zijn gemeten met stroomsnelheidsmeters en waterpeilen met druksensoren die waren geïnstalleerd op de twee locaties. De Chezy ruwheidcoëfficiënt (C) en afvoer coëfficiënten van de kunstwerken (Cd) zijn gebruikt als parameters voor de kalibratie. Voor het opzetten van het model, kanaal bodem profielen en dwarsdoorsneden zijn bemeten met landmeetkundige apparatuur van Total Station. Gedetailleerde gegevens over de locatie en de kenmerken van kunstwerken zijn ook op locatie bepaald. Het hydraulisch functioneren in de huidige situatie, evenals bij beheer scenario's die de operationele efficiëntie, billijkheid en besparing van irrigatiewater zouden moeten verbeteren zijn met het model gesimuleerd.

In aanvulling op de evaluatie van gemeten afvoer gegevens, maakte de hydraulische simulatie van het Metahara systeem een beter begrip van de hydrodynamica en het functioneren van de wateraanvoer bij het huidige beheer mogelijk. Simulatie heeft geresulteerd in een jaarlijks overtollige wateraanvoer van 41 Mm^3 (miljoen kubieke meter), dat is 27% van de jaarlijkse vraag. Het gesimuleerde overschot komt goed overeen met het teveel aan afgevoerd water, dat op basis van routinematige debiet metingen is bepaald op 37 Mm^3. Simulaties hebben ook aangetoond dat het dagelijks vullen en ledigen van de kanalen bij het huidige beheer meer fluctuatie in de waterstanden in de midden sectie veroorzaakt dan in de bovenstroomse en benedenstroomse secties. Vandaar dat de maximale fluctuaties in wateraanvoer bij de inlaatwerken zijn waargenomen in de midden sectie. Er was een snelle afname in de waterstanden gedurende twee tot drie uur na opening van de inlaatwerken. De hydraulische gevoeligheid van de kunstwerken in de midden sectie en onvoldoende beheer van het aflaatwerk van het reservoir en regelwerken voor het waterpeil waren de belangrijkste oorzaken. Anderzijds, namen de afvoeren bij de inlaatwerken in het bovenstroomse deel over het algemeen geleidelijk toe ten gevolge van een toename van de waterstanden in het bovenstroomse kanaal tijdens de irrigatie uren in deze sectie. Debieten naar de benedenstroomse inlaatwerken, bleven echter min of meer hetzelfde tijdens irrigatie uren.

Al met al is gevonden dat de hoeveelheid water die verloren ging op tertiair en veldniveau slechts 7% van het overtollige water was. Wegzijging in de hoofd en secundaire kanalen was relatief klein als gevolg van verminderde infiltratie door verstopping door fijne rivier sedimenten. De simulatie toonde aan dat meer dan 50% van het teveel aangevoerde water bij het stroomafwaartse uiteinde van het systeem werd afgevoerd, waar duidelijk ernstige wateroverlast en verzilting optraden. Met betrekking tot de effectiviteit van de aanvoer, zijn de benedenstroomse inlaatwerken gemiddeld voorzien van een relatieve aangevoerde hoeveelheid van 1,17. Zodra deze inlaatwerken waren gesloten liep het water de benedenstrooms gelegen moerassen in. De bovenstroomse inlaatwerken en die in de midden sectie hadden relatieve aanvoeren van respectievelijk 1,05 en 0,84. De op basis van de gesimuleerde afvoeren bij de inlaatwerken bepaalde gemiddelde operationele efficiëntie van de bovenstroomse, midden en benedenstroomse inlaatwerken, die allemaal goed functioneren onder het huidige beheer, waren respectievelijk 0,93, 0,94 en 0,85. De algemene billijkheid van wateraanvoer naar de inlaatwerken langs het kanaal onder het huidige beheer is bepaald op basis van een ruimtelijke variatiecoëfficiënt (CV) van 0,15, wat als 'redelijk' kan worden beschouwd. Als zodanig zijn de tekortkomingen in het hydraulisch functioneren van het Metahara systeem die op basis van de hydraulische simulatie onder het huidige beheer zijn bepaald: 1. overtollige aanvoer van water; 2. benedenstroomse afvoer die resulteerde in wateroverlast; 3. onvoldoende wateraanvoer naar de inlaatwerken in de midden sectie en overtollige aanvoer bij de benedenstroomse inlaatwerken.

Drie verschillende operationele scenario's die zouden moeten leiden tot meer billijkheid, effectiviteit en waterbesparing zijn gesimuleerd en het effect van elk scenario op het hydraulische functioneren is geëvalueerd. De scenario's waren: 1. toepassing van 24 uur irrigatie met een ongewijzigde wateraanvoer in het systeem, 2. toepassing van 12 uur irrigatie met gewijzigde instelling van de inlaatwerken, 3. toepassing van 9 uur irrigatie met gewijzigde instellingen voor het beheer van het hoofd inlaatwerk, het reservoir en andere inlaatwerken. Toepassing van deze operationele scenario's zou leiden tot een jaarlijkse waterbesparing van respectievelijk 15, 11 en 14%, wat aanzienlijke besparingen zijn voor een oppervlaktewater irrigatiesysteem op basis van zwaartekracht. De operationele efficiëntie was bepaald op hoger dan 0,9 in elk scenario voor de gesimuleerde wateraanvoer bij de inlaatwerken. De algemene billijkheid van de wateraanvoer bij de inlaatwerken, bepaald op basis van de effectieve en geleverde (gesimuleerde) wateraanvoer bij de inlaatwerken, was eveneens vrij voldoende (CV tussen 0.06 en 0.12). De verhouding voor het functioneren (relatieve wateraanvoer) was voor de scenario's 1 en 2 voor elke sectie aangegeven als 'goed' en volgens het ontwerp. Voor scenario 3 was het functioneren echter 'redelijk' voor de bovenstroomse en midden secties, terwijl het 'goed' was voor de benedenstroomse sectie.

Vergelijkende analyse van het functioneren van irrigatiesystemen maakt vergelijking tussen systemen en binnen hetzelfde systeem in de loop van de tijd als middel van het signaleren van veranderingen mogelijk. Onderlinge vergelijking van irrigatiesystemen helpt om resultaten van irrigatie en de algemene effecten op landbouwkundige systemen te vergelijken. Externe indicatoren leveren in principe beperkte informatie over de interne processen van het irrigatiesysteem. Bij de vergelijkende evaluatie van het functioneren is niet de werkelijke numerieke waarde van de indicator belangrijk, maar het relatieve functioneren van het landbouwkundige systeem ten opzichte van andere systemen. Terwijl de analyse van het interne functioneren (van het proces) zich vooral richt op de realisatie van de interne doelstellingen voor het waterbeheer, zoals debiet en de timing van de wateraanvoer, verschaft vergelijkende evaluatie inzicht over hoe productief en efficiënt land en watervoorraden worden ingezet voor de landbouw. De Golgota en Wedecha door de gemeenschap beheerde systemen in deze studie zijn geëvalueerd op basis van drie groepen vergelijkende indicatoren, namelijk de watervoorziening, de landbouwproductie en de fysieke duurzaamheid.

Deze twee systemen variëren met betrekking tot verschillende aspecten, waaronder de bron van het water, de wijze van het verkrijgen van water, het waterbeheer, de grootte van grondbezit, enz. Voor het Golgota systeem is water relatief niet schaars en zijn de boeren verantwoordelijk voor alle aspecten van het waterbeheer zonder enige betrokkenheid van een externe overheidsdienst. Bovendien is het water voor de watergebruikers gratis, met uitzondering van hun eigen routinematige onderhoud. Voor het Wedecha systeem is de beslissing betreffende waterlevering vanuit de bron echter in handen van een externe overheidsdienst, terwijl de boeren zijn verantwoordelijk zijn voor de waterverdeling en het waterbeheer op kavel niveau. Boeren in het Wedecha systeem betalen voor irrigatiewater een bijdrage van 48 US$/ha per jaar aan de overheidsdienst. Op basis van deze verschillen is een vergelijkende evaluatie uitgevoerd om het gebruik van land en water voorraden en de duurzaamheid van de irrigatie te onderzoeken. De twee groepen van indicatoren voor de vergelijking (watervoorziening en landbouwkundige productie) zoals voorgesteld door het *International Water Management Institute (IWMI)* zijn gebruikt, waaraan een derde groep genaamd fysieke duurzaamheid indicatoren is toegevoegd.

De vergelijkende evaluatie van het functioneren liet zien dat er een significant verschil is in het gebruik van de watervoorraden door de systemen. In het Golgota systeem, waar alle aspecten van het waterbeheer de verantwoordelijkheid van de watergebruikers is, was de jaarlijkse relatieve watervoorziening meer dan het dubbele van het Wedecha systeem. Institutionele aspecten voor de aanvoer van water en de bijdrage voor het gebruik van irrigatiewater zijn geïdentificeerd als de belangrijkste factoren voor een efficiënt gebruik van water in deze systemen. Hoewel participatief irrigatiewater beheer in het Wedecha systeem geresulteerd heeft in een verminderde wateraanvoer, hadden de roulatie van de wateraanvoer en mate van betrouwbaarheid hun eigen invloed op de productiviteit. Productiviteit van water was in het Golgota systeem relatief inferieur ten opzichte van het Wedecha systeem. Vanwege het huidige schaarsere water bij het Wedecha systeem, lijkt dit juist. Anderzijds is de extreem lage productiviteit van water in het Golgota systeem echter een zorg, zelfs bij een royale wateraanvoer. De productiviteit van land bleek in het Golgota systeem bijna twee keer zo hoog te zijn als die van het Wedecha systeem. De productiviteit van land is echter niet alleen een functie van de beschikbaarheid van water, maar ook van andere factoren zoals grondsoort, bemesting, plantenrassen, enz. Er is ook gevonden dat de beschikbaarheid van water indirect de productiviteit kan beïnvloeden. Vastgesteld is dat de makkelijke water beschikbaarheid in het Golgota systeem de bereidheid van de boeren vergrootte om meer te investeren in hun stuk land en ook leidde tot toename van de irrigatie intensiteit, al deze factoren leidden tot een hogere productie per oppervlakte eenheid. De jaarlijkse productiviteit van het geïrrigeerd land in het Golgota systeem (zo hoog als 6.000 US$/ha) was 600% hoger dan het gemiddelde in Sub-Sahara Afrika. Hoge irrigatie intensiteit (ongeveer 250%) heeft het grootste aandeel in de hoge productiviteit van het land.

Fysieke duurzaamheid was als een indicator bedoeld voor het bepalen van de duurzaamheid van de geïrrigeerde gebieden en voor het gebruik van het land bij de twee door de gemeenschap beheerde systemen. Zowel de irrigatie verhouding als de duurzaamheid waren voor het Golgota systeem hoger. Het ontbreken van een institutionele structuur voor het waterbeheer en van een beleid voor de bijdrage aan irrigatiewater was de belangrijkste gevonden reden voor uitbreiding van geïrrigeerd land in het Golgota systeem.

De vergelijkende evaluatie van het functioneren in de twee systemen heeft de volgende kernpunten opgeleverd: 1. boeren zijn bereid te betalen voor een minimaal routinematig onderhoud door henzelf, maar niet aan een externe overheidsdienst voor waterbeheer; 2. bereidheid van boeren om te investeren in hun stuk land en dus in de productiviteit van het land is afhankelijk van overeenkomsten voor irrigatiewater beheer; 3. voor dergelijke kleine boeren geldt, hoe groter de omvang van hun land, des te hoger de productiviteit van het land is als gevolg van de bereidheid van de boeren om toevoegingen te gebruiken en fulltime op hun stuk land te werken; 4. de geschiktheid van overeenkomsten voor beheer van irrigatiewater is afhankelijk van het type en de conditie van de bron van het water.

Er kan worden geconcludeerd dat een redelijke bijdrage voor het gebruik van irrigatiewater bij een geschikte hoeveelheid een bruikbare stimulans is voor het verbeteren van de productiviteit van het water. Terwijl alle aspecten van het waterbeheer (rotatie, het gezamenlijk gebruik van water, oplossen van conflicten, routinematig onderhoud, enz.) het beste kunnen worden gedaan door de vereniging van watergebruikers, worden interventies door een externe overheidsdienst, vooral voor het meten en controleren van wateraanvoer, in de huidige situatie aanbevolen. Bovendien is vastgesteld dat collectieve volumetrische benadering (beleid met betrekking tot de bijdrage voor watergebruik) veel beter voor een effectief gebruik van water werkt dan de huidige gebiedsafhankelijke bijdrage zoals die nu voor het Wedecha systeem van toepassing is.

Evaluatie van het interne (proces) functioneren van irrigatiesystemen is bedoeld om de interne processen te analyseren, zoals de hoeveelheid wateraanvoer, de timing ervan, de duur ervan, de betrouwbaarheid van de aanvoer, enz. De reden voor interne evaluatie van het functioneren is om de dienstverlening met betrekking tot irrigatie aan de gebruikers te verbeteren. Evaluatie van interne indicatoren vereist over het algemeen gemeten kwantitatieve gegevens over de wateraanvoer. Aan het bemeten van irrigatiewater wordt in de regel in kleinschalige systemen, vooral in de minst ontwikkelde landen, weinig of geen prioriteit gegeven en dergelijke gegevens zijn daarom doorgaans niet beschikbaar. Daarom zouden afvoer gerelateerde gegevens van dergelijke systemen moeten worden verzameld wanneer dit nodig is. Toch zou interne evaluatie van het functioneren van deze systemen op basis van in het veld gemeten afvoergegevens niet goed tegemoet komen aan de behoeften van kleine boeren. Dit komt omdat in door de gemeenschap beheerde systemen met een slechte irrigatie infrastructuur op hoofdsysteem en veldschaal niveau, water gebruikers over het algemeen andere en diverse criteria voor de evaluatie van de dienstverlening op het gebied van irrigatie hebben, die bij de conventionele methoden niet aan de orde komen. Daarom is een andere benadering van de evaluatie op basis van de percepties van boeren een alternatief.

Het niveau van dienstverlening op het gebied van irrigatie (voorziening) kan worden geëvalueerd vanuit het perspectief van de watergebruikers (de belangrijkste belanghebbenden in de sector) op basis van hun kwalitatieve antwoorden met betrekking tot wateraanvoer. In de Golgota en Wedecha systemen waren geen gegevens over irrigatie water aanvoer en de bijbehorende timing om de dienstverlening aan elke groep van watergebruikers te kunnen evalueren. Het functioneren van de dienstverlening op het gebied van irrigatie is daarom geëvalueerd op basis van kwalitatieve gegevens verzameld bij een steekproef onder de water gebruikers op verschillende locaties binnen de systemen. Drie factoren voor het functioneren, namelijk hanteerbaarheid, timing en betrouwbaarheid, zijn gebruikt en elke factor werd ontleed in twee subfactoren voor het functioneren. De subfactoren die in beschouwing zijn genomen waren grootte van de wateraanvoer en plaats van de wateraanvoer voor hanteerbaarheid, de tijd van het arriveren van het water en de duur van de wateraanvoer voor timing, en kennis van de toekomstige wateraanvoer en zekerheid over de beschikbaarheid voor de betrouwbaarheid. Percepties van de boeren over het belang en de geschiktheid van elke subfactor zijn verzameld met behulp van een enquête in elk systeem. De *Fuzzy set* theorie is gebruikt om de houding van de watergebruikers' te aggregeren voor de bovenstroomse, midden en benedenstroomse secties. De geaggregeerde kwalitatieve beschrijvingen van de boeren zijn vervolgens omgezet in een numerieke indicator voor de niveaus van dienstverlening (functioneren) die liepen van nul tot een.

De resultaten van de analyse van het functioneren zijn bepaald voor zowel het belang van de factoren en de geschiktheid van de dienstverlening ten opzichte van de factoren voor het functioneren. Voor het Golgota systeem was hanteerbaarheid de belangrijkste factor, terwijl betrouwbaarheid de minst belangrijke was. Boeren maakten zich meer zorgen over de hoeveelheid wateraanvoer en het punt van de wateraanvoer dan over de betrouwbaarheid ervan. Anderzijds was voor beide subsystemen van het Wedecha systeem de betrouwbaarheid de belangrijkste factor, terwijl timing de minst belangrijke was, wat aangeeft dat zij zich meer zorgen maakten over de zekerheid en betrouwbaarheid van de beschikbaarheid van water. Voor het Golgota systeem was het totale geaggregeerde functioneren voor de midden sectie hoger, terwijl het hetzelfde en lager was in de bovenstroomse en benedenstroomse secties. Anderzijds nam voor het Wedecha systeem de algemene voorziening van de bovenstroomse sectie naar de benedenstroomse sectie af.

De doelstelling was om het functioneren van de dienstverlening op het gebied van irrigatie te verbeteren en daarmee de productiviteit te verhogen. Daarom is de gemiddelde landbouwkundige productiviteit in elke sectie van beide systemen bepaald teneinde elke relatie met het functioneren na te gaan. De resultaten gaven aan dat gemiddeld bleek dat er in het Golgota systeem geen relatie is tussen het functioneren en resultaat als gevolg van meer uniforme en betere waarden voor het functioneren over de secties. Bovendien, is de landbouwproductie ook een functie van een aantal andere elementen van het landbouwkundige systeem die niet in beschouwing zijn genomen. In het Wedecha systeem is de gemiddelde landbouwkundige productie echter met het functioneren van bovenstrooms tot benedenstrooms gestaag gedaald. De geïdentificeerde aspecten die van fundamenteel belang voor de kwaliteit van de dienstverlening op het gebied van irrigatie (functioneren) zijn: institutionele regeling voor het waterbeheer, aanwezigheid van kunstwerken voor de waterverdeling en op kavelniveau en hun goede functioneren, het realisatie vermogen van de verenigingen van watergebruikers.

Ten slotte zijn de fundamentele verschillen met betrekking tot het functioneren van de grootschalige en kleinschalige irrigatiesystemen geïdentificeerd en geëvalueerd. Voor de grootschalige systemen waren de overtollige wateraanvoer (noodzaak voor waterbesparing), bedreigingen door de stijgende grondwaterstand (wateroverlast) en verzilting, evenals inefficiënte handmatige bediening die resulteerde in slecht hydraulisch functioneren de belangrijkste punten van zorg. De belangrijkste bedreigingen voor de duurzaamheid van deze systemen zijn verzilting en ondiepe grondwaterstanden die het gevolg zijn van overtollige wateraanvoer en een slecht hydraulisch functioneren. Bijna 1.000 ha land in elk systeem, vooral in de benedenstroomse delen, waren onder de dreiging van ondiepe grondwaterstanden en zout grondwater. Deze systemen hebben geen ondergronds drainage systeem voor het controleren van het grondwater peil. Gecontroleerde water inlaat en het verbeteren van het hydraulische functioneren door adequaat beheer bespaart niet alleen aanzienlijke hoeveelheden zoet water, maar vermindert ook het risico van verdere wateroverlast en zorgt voor duurzaamheid. Voor de kleinschalige systemen hielden belangrijke zorgen voor het functioneren verband met de institutionele regelingen voor het waterbeheer, dienstverlening op het gebied van irrigatie, productiviteit van land en water, en waterbeheer op hoofdsysteem en op kavel niveau. Omdat dit systemen voor kleine boeren zijn, is productiviteit (zowel van land als van water) cruciaal voor hen. De duurzaamheid van deze systemen is afhankelijk van de duurzaamheid van het institutionele functioneren voor adequaat water beheer, beheer en onderhoud, en de betrouwbaarheid van de aanvoer van irrigatiewater.

Terwijl het mogelijk is om een vergelijking te maken tussen grootschalige en kleinschalige gemeenschap beheerde irrigatiesystemen, is het ook belangrijk om de fundamentele verschillen te onderkennen. In feite is er een fundamenteel verschil in typologie tussen deze systemen. De grootschalige systemen zijn door de overheid beheerde systemen met een mono cultuur (suikerriet) en geen individuele watergebruikers. Daarom zijn waterbeheer vraagstukken met betrekking tot het verdelen van water onder boeren en de interne dienstverlening niet van toepassing. Suikerriet is een meerjarig gewas met een min of meer uniforme waterbehoefte. In de door de gemeenschap beheerde systemen, oefenen de boeren verschillende teeltsystemen uit voor intensivering. Als gevolg van min of meer vergelijkbare waterbeheer en landbouwpraktijken op de grootschalige systemen, bleef de productiviteit water en land nagenoeg gelijk. Productiviteit van de grond (de productie per ha op basis van de netto-omzet) bij suikerriet zonder verwerking is ongeveer 425 US$/ha per jaar, wat slechts 1/8 is van de land productiviteit van het Golgota systeem en een kwart van die van het Wedecha systeem. Echter, door de verwerking van suikerriet tot suiker steeg de jaarlijkse netto productiviteit van de grond met 550%. Netto productiviteit water in de grootschalige systemen met Suikerriet is ongeveer 0,018 US$/m^3, terwijl die bij de door de gemeenschap beheerde systemen varieert tussen 0,1 en 0,3 US$/m^3 voor de geleverde hoeveelheid irrigatiewater. Verwerking van suikerriet tot suiker leidde inderdaad tot een aanzienlijk hogere productiviteit van water en land.

In het algemeen is adequaat waterbeheer bij geïrrigeerde landbouw van grote betekenis voor de toekomst van de Ethiopische landbouw. De korte termijn ontwikkelingsplannen van het land op het gebied van irrigatie laten zien dat kleinschalige irrigatiesystemen worden beschouwd als de belangrijkste leveranciers van voedsel, terwijl ontwikkelingen met betrekking tot de grootschalige irrigatie systemen grotendeels bestemd zijn voor door de overheid beheerde grootschalige agro verwerkende industrie, voornamelijk suiker. In feite kan het belang van kleinschalige irrigatiesystemen voor de voedselzekerheid in Ethiopië goed worden onderkend als gevolg van de demografische en grondbezit situatie. Kleinschalige landbouw (geïrrigeerd en regenafhankelijk) leveren momenteel meer dan 95% van de voedselproductie. Toch lijkt het noodzakelijk om het landbouw systeem in Ethiopië, dat vooral gebaseerd is op eigen gebruik, te transformeren naar middelgrote en grootschalige geïrrigeerde landbouw voor duurzame voedselzekerheid. Dit is ook belangrijk voor de sector om haar aandeel te hebben in de ontwikkelingsplannen van het land. Als zodanig zou de ontwikkeling van het potentieel voor middelgrote en grootschalige irrigatie in de laagland gebieden voor de productie van voedselgewassen moeten worden versneld. Ondertussen verdient het algemene functioneren en duurzaam beheer van de ontwikkelde irrigatie systemen een gelijkwaardige aandacht. In dit verband zijn het functioneren van de institutionele regelingen, effectiviteit van het fysieke beheer, effectiviteit van het onderhoud, solide roulatie van irrigatiewater voorziening, dienstverlenend beheer en betrouwbaarheid belangrijke aspecten die goed moeten worden geïntegreerd in het beleid op het gebied van het beheer van de water voorraden van het land en de bij de uitvoering betrokken partijen.

Annex J. About the author

Zeleke Agide Dejen was born on the 7th of August, 1978 in Harar, Ethiopia. He attended his primary and secondary education in the small town of Bedessa, West Hararge, Ethiopia. He then joined Arba Minch University, the then Arba Minch Water Technology Institute, in 1996 for his higher education and graduated with a BSc degree in Irrigation Engineering in 2001. From 2001 to 2005, he worked at Arba Minch University as an Assistant Lecturer. His duties were teaching, research and supervising BSc theses preparation of graduating students of Water Resources and Irrigation Engineering.

In 2005, he joined UNESCO-IHE Institute for Water Education for his MSc study in Hydraulic Engineering-Land and Water Development where he finished the study in April 2007 with good academic records. He then returned back to Ethiopia and joined the Water Technology Institute at Arba Minch University for a career in teaching and research. Upon return he was engaged in teaching, research, consultancy and supervision of both BSc and MSc theses of graduating students. He also served as the Dean of the Water Technology Institute for about one and half years (October 2008 to February 2010) during which he was responsible for the overall academic affairs of the Institute which runs three academic departments under it.

In February 2010, after Prof. Bart Schultz had agreed to act as his promoter, he was admitted to the PhD programme in Hydraulic Engineering- Land and Water Development at the UNESCO-IHE Institute for Water Education, in Delft, The Netherlands. During his PhD research period he attended and gave oral presentations at three consecutive annual international conferences of the International Commission on Irrigation and Drainage (ICID). These conferences were in Tehran, Iran (2011); Adelaide, Australia (2012); and Mardin, Turkey (2013). He also made a presentation at the 14th Annual Symposium on Sustainable Water Resources Development at Arba Minch University in June 1014. All his conference presentations were related to his PhD research. He published two articles related to his PhD research in international peer-reviewed journals. In addition, one more paper is under preparation. Moreover, he presented the progress of his PhD research at four of the Annual PhD seminars of UNESCO-IHE held every year in September. UNESCO-HE, being part of the Research School for Socio-Economic and Natural Sciences of the Environment (SENSE), he attended several PhD courses within the SENSE network.

T - #1022 - 101024 - C188 - 240/170/10 - PB - 9781138027671 - Gloss Lamination